Joachim Klima

Lubricant transport towards tribocontact in capillary surface structures

Karlsruher Institut für Technologie
Schriftenreihe des Instituts für Technische Mechanik

Band 31

Eine Übersicht aller bisher in dieser Schriftenreihe erschienenen Bände finden Sie am Ende des Buchs.

Lubricant transport towards tribocontact in capillary surface structures

by
Joachim Klima

KIT Scientific Publishing

Dissertation, Karlsruher Institut für Technologie
KIT-Fakultät für Maschinenbau

Tag der mündlichen Prüfung: 15. März 2018
Referenten: Prof. Dr.-Ing. Wolfgang Seemann, Prof. Dr. Nicolas Fillot

Impressum

KIT Scientific Publishing

Karlsruher Institut für Technologie (KIT)
KIT Scientific Publishing
Straße am Forum 2
D-76131 Karlsruhe

KIT Scientific Publishing is a registered trademark
of Karlsruhe Institute of Technology.
Reprint using the book cover is not allowed.

www.ksp.kit.edu

Print on Demand 2018 – Gedruckt auf FSC-zertifiziertem Papier

ISSN 1614-3914
ISBN 978-3-7315-0814-4
DOI 10.5445/KSP/1000084131

Lubricant transport towards tribocontact in capillary surface structures

Zur Erlangung des akademischen Grades
Doktor der Ingenieurwissenschaften

der KIT-Fakultät für Maschinenbau
Karlsruher Institut für Technologie (KIT)

genehmigte
DISSERTATION

von

M.Sc. Joachim Klima

Tag der mündlichen Prüfung:	15. März 2018
Hauptreferent:	Prof. Dr.-Ing. Wolfgang Seemann
Korreferent:	Prof. Dr. Nicolas Fillot

Acknowledgements

The present work would not have been possible without the help and support of many people: First of all, I wish to express my deep and sincere gratitude to my advisor Prof. Dr.-Ing. Wolfgang Seemann for his continuing guidance and support – with respect to small details and regarding the bigger picture. Both were necessary especially due to the scarcity of closely related research. He gave me a lot of freedom to pursue my own thoughts, but in the decisive moments he managed to keep me on track.

I am indebted to Prof. Dr. Nicolas Fillot for his input and valuable discussions. He especially helped me to look beyond my narrow research topic and see further links within the vast field of tribology. For challenging me with critical comments and discussions I want to thank Prof. Dr.-Ing. habil. Alexander Fidlin. This repeatedly helped me in refining my thinking.

Furthermore, I would like to express my gratitude to Prof. Dr.-Ing. Hartmut Hetzler for inspiring discussions particularly in the initial phase of my work. I really enjoyed my time at the Institute of Engineering Mechanics in Karlsruhe interacting with an amazing group of smart, welcoming, humble, and supportive individuals – thank you.

At Robert Bosch GmbH, which funded this research, I owe my advisors, Dr. rer. nat. Gerd Dornhöfer and Dr. rer. nat. Markus Hinterkausen, my deepest gratitude. First of all, it was their initial idea that my whole effort is based on. With great knowledge and experience, they guided me each in his individual way to make me and the work advance.

Numerous fantastic colleagues at Bosch Corporate Research shared their time, thoughts, equipment, and expertise or simply laughed with me – it was hard work and great fun. My special thanks for the insightful discussions we had as well as their dedication go to M.Sc. Sigrid Herb, M.Sc. Lei Zhang, M.Sc. Florian Bühner, and M.Sc. Dmitriy Mikhaylov.

For cooperation, support, and inspiration I am furthermore indebted to the laser experts at KIT IAM, the Bosch colleagues in Schwäbisch Gmünd as well as researchers at INSA, Lyon.

Finally, a life (and working on this research was a considerable part of my life these past years) is nothing without friends and family, whom I would deeply like to thank: above all my parents and Maria.

Abstract

Greater motivation for this work is to counter lubricant shortage at a frictional contact where two bodies in relative motion meet. Lubrication generally aims at improving frictional behavior. If there is not enough lubricant available for the tribocontact, so-called starvation occurs which may lead to increased friction, wear, and related undesired effects. In the worst case this results in total failure of the respective machine or assembly.

The specific idea is to transport lubrication liquids, for example oils, from a distant location towards an undersupplied tribocontact in small channels or generally structures cut into a flat surface. Concentration is on the transportation process with the tribocontact itself playing a minor role. In addition to guiding liquids, microscale surface structures allow to use capillary force as a cheap and reliable driver of the flow. To make the best use of it, different surface designs are discussed resulting in the microchannel as advantageous basic building block.

Existing microchannel flow theory is adapted and extended to meet typical conditions in the vicinity of a tribocontact. Fabrication restrictions due to the materials in use and resulting consequences are covered. Inertial effects from moving solid bodies which carry the structure are discussed and thermal effects which are caused by the heat tribocontact generates are equally considered. If sufficiently strong, both may hinder or even suppress the lubricant supply of a tribocontact. This is countered by capillary force in suitable surface geometries.

Starting from basic fluid mechanics laws, a nondimensionalization and approximation procedure leads to significant simplification in the governing equations which results at last in a computationally inexpensive unidirectional flow model. Special attention is paid to the boundaries of the fluid domain in order to take non-standard solid geometries and an important interfacial phenomenon at the liquid-air boundary – the so-called Marangoni effect – into account. Numerical modeling is complemented and validated by flow experiments on specially designed testing equipment.

Even though the major focus of this work is on the investigation of liquid lubricant transport towards tribocontact, the transition from such transportation domain into the tribocontact itself cannot be neglected as it proves important for overall system functioning. Various potential transition mechanisms are discussed including tribocontact relative motion and starting points for future research are given – with respect to both physical insight and the harmonization of basic models in the respective domains. In general, room is left for further research both in numerical modeling and solution techniques as well as in more comprehensive experimental assessment.

Zusammenfassung

Übergeordnete Motivation dieser Arbeit ist die Bekämpfung des Schmierstoffmangels an einem Reibkontakt – dort, wo sich zwei Körper berühren und sich relativ zueinander bewegen. Geschmiert wird generell um das Reibverhalten zu verbessern. Steht dem Tribokontakt nicht ausreichend Schmierstoff zur Verfügung, tritt die sogenannte Mangelschmierung auf, die zu unerwünschten Effekten wie erhöhter Reibung und erhöhtem Verschleiß führen kann. Im schlimmsten Fall kommt es zum Totalausfall der betroffenen Maschine oder Baugruppe.

Die konkrete Idee ist nun, flüssige Schmierstoffe, z.B. Öle, aus entfernten Bereichen zum unterversorgten Tribokontakt zu transportieren und dafür bestimmte Strukturen – beispielsweise Kanäle – zu nutzen, die in eine vormals plane Oberfläche eingebracht wurden. Im Mittelpunkt steht dabei der Transport an sich, während dem Tribokontakt eine untergeordnete Rolle zukommt. Die genannten Mikrostrukturen können – zusätzlich zu ihrer Funktion, Flüssigkeiten zu leiten – Kapillarkräfte entstehen lassen, die dann dazu genutzt werden können, Strömungen anzutreiben und zwar auf kostengünstige und zuverlässige Art und Weise. Um dies bestmöglich auszunutzen, werden unterschiedliche Oberflächenstrukturen diskutiert; der Mikrokanal als Grundbaustein einer Oberfläche stellt sich dabei als vorteilhaft heraus.

Die bestehende Theorie zu Mikrokanalströmungen wird erweitert und an die typischen Bedingungen in der Umgebung eines Tribokontaktes angepasst. Fertigungsrestriktionen aufgrund verwendeter Materialien und daraus entstehende Konsequenzen werden betrachtet. Auftretende Trägheitskräfte, weil sich die strukturierten Bauteile bewegen, werden ebenso berücksichtigt wie thermische Effekte, die der Wärmeentwicklung im Tribokontakt geschuldet sind. Beides kann dazu führen, dass die Versorgung des Tribokontakts mit Schmierstoff behindert oder sogar gänzlich unterbunden wird. Die Kapillarkraft wirkt dem bei geeignetem Oberflächendesign entgegen.

Grundlegende Gleichungen der Fluidmechanik werden mittels Entdimensionalisierung und anschließender Näherung zu einem unidirektionalen Strömungsmodell vereinfacht, das schnelles Rechnen erlaubt. Besondere Beachtung finden dabei die Ränder des Fluidgebiets, sowohl hinsichtlich spezieller Strukturgeometrien als auch im Hinblick auf ein bedeutendes Grenzflächenphänomen zwischen Schmierflüssigkeit und Luft, dem sogenannten Marangoni-Effekt. Die angestellten Berechnungen werden durch Experimente auf einem eigens entwickelten Versuchsaufbau ergänzt und validiert.

Auch wenn das Hauptaugenmerk dieser Arbeit auf der Untersuchung des Schmierstofftransports hin zum Tribokontakt liegt, darf nicht außer Acht gelassen werden, dass der Übergang vom Transportgebiet in den eigentlichen Tribokontakt für die Funktionsweise des Gesamtsystems von besonderer Bedeutung ist. Mehrere potentielle Übergabemechanismen, u.a. Relativbewegungen im Tribokontakt, werden diskutiert und Ausgangspunkte für zukünftige Forschungen werden beschrieben – sowohl was das grundlegende physikalische Verständnis als auch die Zusammenführung bestehender Modelle angeht. Generell besteht Spielraum für weitere Forschungen sowohl im Bereich weiter gehender Berechnungsmodelle und Lösungsverfahren als auch was umfassendere Versuchsreihen angeht.

Contents

Nomenclature

Acronyms

ABS	Acrylonitrile butadiene styrene
ASTM	ASTM International
DIN	Deutsches Institut für Normung
DRIE	Deep reactive ion etching
EHD	Elastohydrodynamic(s)
FVA	Forschungsvereinigung Antriebstechnik
ISO	International Organization for Standardization
LIGA	Lithographie, Galvanoformung, Abformung
NIST	National Institute of Standards and Technology
NTP	Normal temperature and pressure
PAO	Polyalphaolefin
PIV	Particle image velocimetry
PMMA	Polymethylmethacrylate
POM	Polyoxymethylene
PTFE	Polytetrafluoroethylene
SEM	Scanning electron microscope
SOR	Successive over-relaxation
VG	Viscosity grade

Dimensionless numbers

Bo	Bond number	[-]
Ca	Capillary number	[-]
Kn	Knudsen number	[-]
M	Mach number	[-]
Ma	Marangoni number	[-]
Nu	Nusselt number	[-]
$\mathrm{Pé}_{\mathrm{th}}$	thermal Péclet number	[-]

Po	Poiseuille number	[-]
Re	Reynolds number	[-]
$\mathrm{Re_r}$	reduced Reynolds number	[-]

Greek symbols

$\boldsymbol{\alpha}$	Angular acceleration	[rad/s^2]
β_T	Isothermal compressibility	[1/Pa]
Γ	Operator used in meniscus surface computation	[-]
γ, γ_{la}	Surface tension at liquid-air interface	[N/m]
γ_{sa}	Interfacial tension at solid-air interface	[N/m]
γ_{sl}	Interfacial tension at solid-liquid interface	[N/m]
Δp	Pressure difference	[Pa]
Δp_f	Pressure drop across front meniscus	[Pa]
Δp_u	Pressure drop across upper meniscus	[Pa]
δ	Ratio of characteristic height h_0 to characteristic length l_0	[-]
ϵ	Average roughness height	[m]
ϵ_{rel}	Relative roughness	[-]
ζ	Auxiliary integration variable	[m]
η	Dynamic viscosity	[Pa s]
η_{avg}	Dynamic viscosity, averaged across film height	[Pa s]
θ	Contact angle	[°]
θ_{app}	Apparent contact angle	[°]
θ_a, θ_r	Advancing, receding contact angles	[°]
θ_{eq}	Equilibrium contact angle	[°]
κ_D	Permeability of a medium	[m^2]
$\varkappa$	Dimensionless quantity in non-inertial frame analysis	[-]
Λ	Channel aspect ratio	[-]
λ, λ_l	Thermal conductivity of liquid	[W/(m K)]
λ_s	Thermal conductivity of solid	[W/(m K)]
μ	Mean value	[..]
ν	Kinematic viscosity	[m^2/s]
ξ	Dimensionless quantity in non-inertial frame analysis	[-]
ρ	Density	[kg/m^3]
$\boldsymbol{\sigma}$	Cauchy stress tensor	[Pa]

ς	Standard deviation	[..]
$\boldsymbol{\tau}$	Stress deviator tensor	[Pa]
τ	Shear stress	[Pa]
τ_M	Shear stress due to Marangoni effect	[Pa]
$\dot{\Upsilon}$	Shear rate	[1/s]
ϕ	Solid corner angle	[°]
φ	Half edge angle	[°]
χ	Thermal diffusivity	[m^2/s]
Ω_c	Corner flow region	[-]
Ω_f	Front meniscus region	[-]
Ω_u	Bulk flow region	[-]
ω,$\boldsymbol{\omega}$	Angular velocity	[rad/s]

Roman symbols

A	Channel cross-sectional area	[m^2]
A^I	Interfacial area	[m^2]
A_u^I	Area of fluid column's top surface	[m^2]
A_S	Filled area within channel cross-section	[m^2]
$\boldsymbol{a}_{C,E,Z,T}$	Coriolis, Euler, centripetal, translational acceleration components	[m/s^2]
B_λ	Coefficient in T-dependent thermal conductivity computation	[1/K]
B_{cp}	Coefficient in T-dependent specific heat cap. computation	[W/(m K^2)]
B_T	Auxiliary term in temperature field computation	[1/m]
b	Channel width or channel diameter	[m]
b_c	Constricted channel width or constricted channel diameter	[m]
b_H	Channel width at $z = H$	[m]
b_{tp}	Test piece width	[m]
C	Spatially constant flow driving factor	[1/(s m)]
c_j	Flow characterization constants or factors	[..]
c_p	Specific heat capacity at constant pressure	[J/(kg K)]
c_v	Coefficient of variation	[-]
d_H	Hydraulic diameter	[m]
$\boldsymbol{E}$	Strain rate tensor	[1/s]
$\hat{\boldsymbol{e}}_x, \hat{\boldsymbol{e}}_y, \hat{\boldsymbol{e}}_z$	Unit vectors in Cartesian coordinate system	[m]
F	Force	[N]

F_{cap}	Capillary force	[N]
F_M	Force due to Marangoni effect	[N]
$\boldsymbol{f}$	Total force density	[N/m^3]
$\boldsymbol{f}_b$	Body force density (excluding inertial contribution)	[N/m^3]
$\boldsymbol{f}_C$	Force density component due to Coriolis acceleration	[N/m^3]
f^D	Darcy friction factor	[-]
$\boldsymbol{f}_E$	Force density component due to Euler acceleration	[N/m^3]
$\boldsymbol{f}_f$	Force density due to fluid inertia	[N/m^3]
f_n^F	Coefficients in Fourier series	[m/s]
$\boldsymbol{f}_T$	Force density component due to translational acceleration	[N/m^3]
$\boldsymbol{f}_Z$	Force density component due to centripetal acceleration	[N/m^3]
G	Characteristic channel shape constant	[-]
G_{em}	Theoretical value for G based on free energy minimization	[-]
G_{ex}	Experimentally determined value for G	[-]
G^I	Gibbs free energy	[J]
g	Gravitational acceleration	[m/s^2]
$\breve{\boldsymbol{g}}$	Vector as defined in figure 2.6	[m]
g_n^F	Auxiliary Fourier series term	[1/m]
H	Height of fluid column or film at central location ($y = 0$)	[m]
H^M	Mean curvature	[1/m]
H_w	Height of fluid column at solid boundary	[m]
H_y	Height of fluid column dependent on lateral coordinate y	[m]
h	Channel height	[m]
h_c	Constricted channel height	[m]
h^{sa}, h^{sl}	Solid-air, solid-liquid heat transfer coefficients	[W/(m^2 K)]
h_{tp}	Test piece height	[m]
(I)	Inertial frame of reference	[-]
J_H	Hydraulic resistance	[1/s]
$K_{Len}^{(\square)}$	Correction factor for channel shape constant	[-]
k_n^F	Fourier coefficients	[-]
L	Length of fluid column	[m]
L_c	Length of liquid fingers	[m]
L_f	Axial length of front meniscus	[m]

L_u	Length of bulk fluid column	[m]
l	Channel length	[m]
l_{cap}	Capillary length	[m]
l_d	Lattice spacing	[m]
l_r	Reservoir length	[m]
l_{tp}	Test piece length	[m]
M	Origin of an auxiliary coordinate system in the reference frame (O)	[-]
$\boldsymbol{n}$	Unit normal vector	[-]
(O)	Non-inertial frame of reference	[-]
p	Pressure	[Pa]
$\breve{\boldsymbol{p}}$	Vector as defined in figure 2.6	[m]
p_{am}	Ambient pressure	[Pa]
p_i,p_o	Inlet pressure, outlet pressure	[Pa]
p_m	Characteristic cross-sectional pressure	[Pa]
Q	Normalized volumetric flow rate	[-]
Q_{th}	Heat generated by external sources	[W/m^3]
q	Volumetric flow rate	[m^3/s]
$\breve{\boldsymbol{q}}$	Vector as defined in figure 2.6	[m]
q_{cr}	Critical flow rate	[m^3/s]
R_c	Non-straight microchannel radius of curvature	[m]
R_u	Upper meniscus cross-sectional radius of curvature	[m]
$R_{1,2}$	Principal radii of curvature	[m]
r	Circular channel or tube radius	[m]
$\breve{\boldsymbol{r}}$	Vector as defined in figure 2.6	[m]
S	Cross-sectional channel saturation	[-]
S^F	Simplified rectangular function	[1/(s m)]
$\breve{\boldsymbol{s}}$	Vector as defined in figure 2.6	[m]
$s_o^{\perp}$	Open boundary length in cross-sectional plane	[m]
$s_w^{\perp}$	Wetted perimeter in cross-sectional plane	[m]
T	Temperature	[K]
T_a	Ambient temperature	[K]
T_{tf}, T_{tb}	Temperature at the test piece's front and back side, respectively	[K]
T_w	Solid body temperature	[K]

t	Time	[s]
$\boldsymbol{t}_1$	Unit tangent vector, in principal direction of curvature maximum	[-]
$\boldsymbol{t}_2$	Unit tangent vector, in principal direction of curvature minimum	[-]
U	Meniscus surface	[m]
$\boldsymbol{u}$	Fluid velocity vector	[m/s]
u,v,w	Fluid velocity in downstream, lateral, vertical direction	[m/s]
$\boldsymbol{u}_{A,B}$	Surface velocity of solid body A, B, respectively	[m/s]
u_h	Downstream fluid velocity at the upper fluid surface	[m/s]
u_w	Downstream fluid velocity adjacent to a solid wall	[m/s]
V	Fluid volume	[m^3]
W	Washburn flow behavior factor	[m^2/s]
X_f	Downstream location where cross-sectional meniscus becomes flat	[m]
X_i,Y_i,Z_i	Distinct locations on respective coordinate axis	[m]
$\boldsymbol{x}$	Position vector of fluid parcel	[m]
x,y,z	Longitudinal (downstream), lateral, and vertical channel coordinates	[m]
x_{cg}	Downstream location of fluid column's center of gravity	[m]
y_r	Radial coordinate in circular tube	[m]
$y_{0(mc)}$	Characteristic microchannel width	[m]
$y_{0(tf)}$	Characteristic width in thin film theory	[m]

Sub- and superscripts, further symbols

∇_S	Surface gradient	[-]
$()_{A,B}$	With respect to body or phase A, B, respectively	
$_I(), {}^I()$	With respect to inertial reference frame	
$()_n$	Normal temperature (20°C) and/or pressure (1 atm)	
$_O(), {}^O()$	With respect to non-inertial reference frame	
$()_t$	With respect to time	
$()_{x,y,z}$	Component acting in the respective coordinate direction	
$()_0$	Characteristic quantity	
$()^*$	Dimensionless quantity	
$()^{(\circ),(\square),(3\varsigma)}$	With respect to circular, square, 3ς Gaussian cross-sectional channel shapes	
$\bar{()}$	Cross-sectional average	
$\tilde{()}$	Based on constant saturation assumption	

1. Introduction

1.1. Motivation and scope

Friction is both a blessing and a curse: Whereas in preventing wheelspin it only allows us to move our car on the road, it uses up about one third of the same car's fuel energy [101]. Lubrication is amongst tribologists' best means to reduce frictional losses in any machinery or product. Key property of lubrication fluids is their ability to significantly lower friction by separating otherwise contacting solid bodies from each other, at least partially. For this to work, enough lubricant must be present at tribocontact in order to prevent the occurrence of starvation. In the case of starvation typical consequences are higher friction and wear. This is generally attributed to lower lubricant film height in tribocontact. A sufficiently high degree of starvation may result in the failure of the tribological system and possibly of the whole mechanical system i.e., of a machine or at least an assembly or machine element. While in general starved lubrication has to be avoided, the phenomenon itself is still not fully understood. Active research concentrates on the effects starvation has on tribocontact behavior. The cause of starvation, in today's understanding an insufficient amount of lubricant at the inlet of tribocontact [39, 48], is addressed in current research to a much lesser extent. The underlying idea is simple: If more lubricant is needed at tribocontact inlet, why not just supply more?

Several mechanisms to transport a lubricant towards tribocontact have been investigated e.g., pumping or spraying. This work concentrates on capillary force as the driver of lubricant transport: Without the need of moving parts or additional equipment, it promises to be a cheap alternative. The general feasibility and reliability of capillary transport is demonstrated in plants – it is their method of choice to transport water and nutrients. The geometrical features required for a capillary force to develop are created by surface structuring i.e., the purposeful deviation from a flat surface, for example by generating holes, grooves, ridges or channels with concentration on the latter. If structure size is limited to the micrometer range or smaller, capillary force is strong in comparison with e.g., gravity. This yet limits the spectrum of potential structuring technologies but still allows for mechanical cutting or laser ablation.

As will be developed in the course of this work, critical attention must be paid to forces that may oppose capillary transport, e.g., a centrifugal force which arises from rotation of a contact body. Besides that, the so-called Marangoni force may hinder or even suppress capillary transport totally. This is a surface force which is present under a thermal gradient like the one that develops when heat is generated in a tribocontact and conducted via solids and fluids to its surroundings. Furthermore, highly stressed tribocontacts require carefully selected engineering materials. These in turn need specific microfabrication processes resulting in consequences for microchannel transport itself which must be considered. While the source of the lubricating fluid is not in the center of attention, it cannot be excluded from the analysis either. Possible liquid lubricant reservoirs can be an oil-filled surface cavity or a grease which – similar to a sponge – is soaked with the oil intended to lubricate tribocontact.

In summary, this research should lay the experimental foundation of the general transport problem as well as build a basic modeling framework which in the optimal case allows fast and reliable numerical simulation. Yet it can only be understood as a first step towards a deeper and more general understanding of the physical processes involved with liquid lubricant supply of a tribocontact out of a distant location.

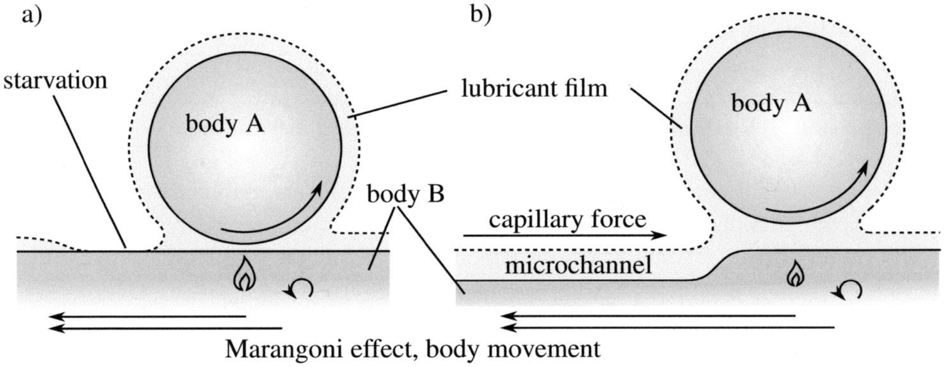

Figure 1.1.: Tribocontact a) in starvation and b) with capillary lubricant supply by a microchannel; thermal effects (heat is generated in the tribocontact) and body movement may draw the lubricant away from the tribocontact with the capillary effect as a countermeasure

1.2. Literature survey

Capillary transport in surface structures has not yet received considerable research attention with respect to the lubricant supply of a tribocontact out of a distant reservoir. Tribology related literature is therefore only part of this overview. Another part deals with general small scale fluid transport, especially capillary transport, which can be applied well to liquid lubricants' movements. This section is designed to give a broad and comprehensive overview of potentially relevant research topics. The essential basics for this work are then dealt with in more detail in the following chapter two.

Methods of lubricant transport towards tribocontact

Active pumping of lubricating fluids is a common method used for example in externally pressurized fluid film bearings but comes with the cost for a pump or at least a pumping mechanism. Spraying of lubricant contained in an aerosol, a well-known technique e.g., in production engineering, recently received more research attention as a means to counter starvation [155]. If a lubricant flow already exists, mechanical sliders can be designed to direct the flow to tribocontact inlet [3]. Van Zoelen et al. [197] have investigated the ability of surface curvature of bearing raceways to transport a lubricant from the side into the center of the raceway i.e., back to tribocontact inlet. This method is interesting as it makes use of existing surface forces which can (under certain circumstances) move oil from one location to another. Insofar it is similar to the present work. Klien et al. [117] have analyzed lubricant migration on solid surfaces with surface roughness and general characteristics typical to tribological conditions based on the so-called Marangoni effect, which is explained in section 2.4.5. They attribute their difficulties in achieving clearly distinguishable results to the presence of capillary forces. The latter are the physical basis of recent bioinspired research by Comanns et al. [42] who investigate the skin structure of moisture-harvesting lizards with the aim in mind to adapt them to (cooling) lubricant transport. They mention the supply of cutting tool edges as well as bearing lubrication as potential application fields. In this respect, laser surface structuring and spreading tests on different surface types have been conducted by Hermens et al. [98].

In precision mechanics, especially in watchmaking, capillary effects are exploited not to transport lubricants towards tribocontact but at least to keep them there. This can be achieved geometrically or by chemically treating surfaces in an epilamization process [157, 199].

Flow within tribocontact

Compared to lubricant flow towards tribocontact, there is abundant literature dealing with flow behavior within contact. Based on Reynolds equation as its mathematical cornerstone, elastohydrodynamic lubrication (EHL) theory provides the toolset to compute not only flow in a tribocontact but also determines lubricant film height between the contact partners, the pressure buildup in the film, the resulting load-carrying capacity of the tribocontact as well as contact partners' elastic deformations, cf. [92, 183]. Extensions of EHL theory to include thermal and surface roughness effects exist [9]. While Reynolds equation itself is generally accepted, the variety of boundary conditions in use e.g., Swift-Stieber, Prandtl-Hopkins, Coyne-Elrod, or Jakobsson-Floberg-Olsson shows that the physics is difficult to model in detail and still is not completely understood, especially towards the edge of tribocontact and towards low pressure regions. How fluid gets entrained into, leaves, and flows around contact is a current research topic both experimentally and theoretically [143, 184, 195] with the latter typically building on roll coating theory [79, 206]. What seems clear is the appearance of starvation effects in case of insufficient lubricant supply at tribocontact's inlet:

Starvation

Starvation effects in elastohydrodynamic lubrication have been investigated for decades in experiments and in theory. Starvation can occur both in oil as well as in grease lubrication [133]. Wedeven et al. [205] point out that starvation is not necessarily always negative. It is rather desirable to understand and control starvation conditions. For this purpose, Cann et al. [32] define five parameters, namely load, speed, oil viscosity, oil volume, and contact size. It is generally agreed that the state of a tribocontact's inlet region is decisive for what happens in the contact, especially with respect to film thickness. While initially, the location of the inlet lubricant meniscus was seen as the critical parameter to control film height [25, 205], later research concentrated on the amount of oil present at the inlet [39, 133]. Such amount is a consequence of the balance between lubricant loss and lubricant replenishment [48]. With respect to the latter, Jacod et al. [104] distinguish between "out of contact" replenishment as mentioned above and "in contact" reflows in or around tribocontact due to local capillary forces.

Surface structuring in tribology

An analogous distinction between "in contact" and "out of contact" can be made when it comes to surface structuring: While the latter is in the center of this works' research, the former is an established research area within the field of tribology: Theoretically based on EHL theory, the goal of surface structuring is to alter contact surfaces within a tribocontact in a way that enhances film formation, pressure generation and thus load-carrying capacity and wear reduction. Key is the basic physical effect that a fluid which is entrained in a converging or suddenly narrowing gap builds up pressure. Etsion [66] gives an overview of both theory and application examples. Figure 1.2a) depicts the pressure increase achieved by a protrusion in contact. Total pressure depends on relative velocity u. An analogous pressure buildup is generated by the single step geometry b) which is the basic building block for multiple step geometries such as c). A variety of surface patterns has been summarized by Wahl [201] including parallel and crossed grooves in different angles, dimples, and protrusions with tribological testing done e.g., by Geiger et al. [80]. A comparison of numerical EHL computations and experiments can be found in [221].

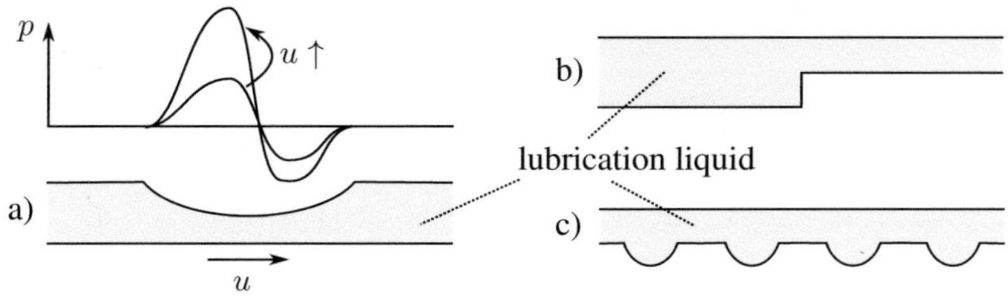

Figure 1.2.: Pressure generation within the tribocontact due to a) a protrusion, b) a step, c) dimples in the surface [66]

While these structures and the ones investigated in the present work share the same ultimate goal, namely the improvement of tribocontact performance, they do it in a significantly different way: The former are located within tribocontact and are not focused on moving lubricants whereas the latter are located in the vicinity of a tribocontact and are specially designed to transport a lubrication fluid from a distant reservoir towards tribocontact inlet.

Microstructure fabrication for fluid transport applications

Outside tribology, there is a large research community dealing with the fabrication of microstructures to transport fluids on a small scale. Typical applications are found in life sciences, chemistry, and environmental science with a special emphasis on so-called "lab-on-a-chip" applications. In electronics, heat-pipes are used for electrical component and microchip cooling [88] while in the field of mechanical engineering, ink jet printing is a common application [141].

The specific field of application often determines or at least bases restrictions on the liquids used and the material that bears the structure with a small overview given e.g., by [189]. As a direct consequence a variety of fabrication methods has been developed to produce surface structures, cf. section 3.2, that range from simple cutting of paper [82], over honing e.g., of an combustion engine's metal or metal-based cylinder surface [57], microinjection molding of engineering plastics [126, 213], etching of silicon and certain polymers [193], mechanical machining by milling or cutting [105] to different variants of laser ablation which are comparable to methods used for in-tribocontact structuring [62].

Microscale fluid transport

The theoretical foundation for fluid transport on the microscale is provided by a field that has been termed "microfluidics" as late as in the 1980s [41]. Not so much the absolute scale of geometrical features (which are roughly in the range between one micron and one millimeter) but the scale's typical physical consequences distinguishes microfluidics research from classical fluid mechanics on one side and nanofluidics on the other.

The basic theory is summarized in e.g., [27, 115] while [125] goes more into detail with respect to fluid transport. Squires and Quake [178] as well as Stone et al. [180] probably give the broadest overview of research in the field, including possibilities of driving a flow e.g., by chemical surface manipulation, thermal gradients, optowetting, electrokinetic techniques, acoustic streaming, and mechanical pumping amongst others. Darhuber and Troian [50] summarize research with respect to microfluidic flow due to modulation of surface stresses. Generally, when surface forces are investigated, the interplay of different phases is crucial. Examples for two-phase systems with an interface between a gaseous and a liquid phase are free drops, liquid jets and related phenomena like the breakup of the latter into the former. The presence of a solid material leads to a three-phase system and is necessary yet not sufficient for capillary forces to occur.

Thin films and droplets

Fluid transport in three-phase systems can be achieved on flat and non-flat surfaces, with the liquid being present only locally such as in channels or in the form of drops or alternatively covering the whole surface as a (thin) film. With respect to the latter, an overview of mathematical description and related physical phenomena which might occur is given e.g., in [147, 149]. Film theory on a structured surface is described in [2]. If the film extends indefinitely, fluid movement cannot be caused by capillarity so that other effects become relevant. Substrate curvature is investigated in [170, 174]. Abundant literature exists in the case of rotating films, mostly related to the so-called "spin-coating" process where the coating is spread on the part by setting it in rotary motion [65, 100, 149]. Temperature variations can lead to movement and instabilities in a thin fluid layer: Rayleigh-Bénard convection is based on density differences in hot and cold sections in the fluid which, under the influence of gravity, give rise to a special movement pattern [12]. Bénard-Marangoni convection or thermocapillary convection is a similar effect which in contrast is caused by differences in surface tension [20, 198]. Alternatively, movement due to the Marangoni effect (which is described in more detail in section 2.4.5) may be caused by chemical agents, so-called surfactants [55]. The effect is also able to pull a liquid layer out of a reservoir [68, 135, 145].

If the fluid volume on a solid surface is finite, the three-phase boundary is characterized by a defined contact angle as described in section 2.2.3. Depending on fluid amount, puddles, drops or droplets form. For the symmetry of interfacial forces, cf. figure 2.4a), they still do not move in the basic case. Methods to move droplets [69, 192] and elongated fluid columns [49] over a temperature-controlled substrate are described as well as motion achieved by locally heating droplet surface with a directed light or laser source [78, 120]. Even droplet motion on a layer of a different fluid is analogously possible [84]. Treating the solid surface chemically to make drops move [21], an effect Wasan et al. [203] more formally set in contrast with the above, directly leads over to the spreading of drops and films due to solid surface characteristics:

Capillarity and spreading on textured surfaces

With the typical neglection of gravity in microscale fluidics, cf. section 2.1, capillary forces are dominant. A broad overview of wetting and spreading behavior in general can be found in [53, 188]. The involvement of capillary effects is a great advantage, because no external energy source is required to make fluids move. Capillary forces depend on geometry and a special surface texture may enhance or hinder wetting. The wetting behavior of such textures has been broadly investigated e.g., in [37, 45, 162]. It is related to fluid transport in porous media, cf. [11]. However, it happens only at the solid's surface. By appropriate structure design, a preferential wetting direction can be achieved [44, 107, 211].

Basic building blocks of a structured surface

As surface geometry often is sufficiently complex with respect to capillary fluid transport, it has been broken down by researchers into basic geometrical features [128]. One of such basic features is a fiber along which a characteristic capillary meniscus develops as seen in figure 1.3a); capillary transport between fibers is then typically described based on the Princen model [17, 159]. Other basic building blocks are tubes and channels:

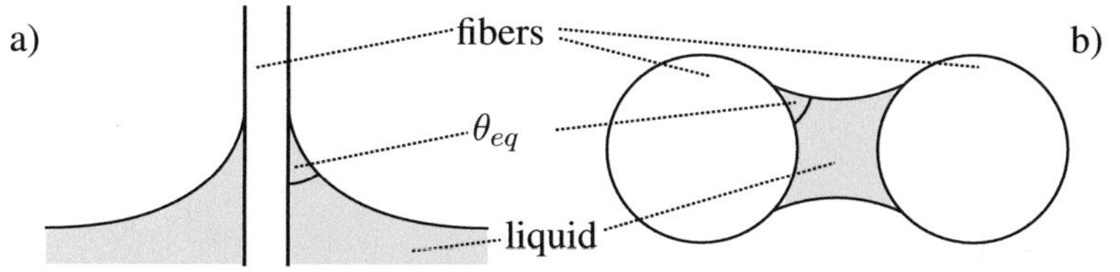

Figure 1.3.: a) Meniscus at a fiber immersed in a liquid with equilibrium contact angle θ_{eq} [53]; b) liquid between two fibers [159]

Flow in tubes and channels

Flow in tubes and channels has been investigated both capillarity-driven, cf. section 2.5, and with other driving mechanisms, most simply by using an external pump. A typical experimental setup is presented in [218] while [1] designed a micro-balloon pump on a microfluidic system. In analogy to droplet transport on a surface, droplet movement inside channels may be achieved by thermal actuation [173]. [127] designed an electrochemical fluid driving method based on the Marangoni effect.

Centrifugation is a further alternative: Rotating microchannels have been investigated mainly in the context of so called "Lab-on-a-CD" systems: Rotating disks are used to transport, handle, mix, valve, or portion fluids on the small scale, again with concentration on applications in medical and biological fields [64, 130, 168, 216]. Capillary effects are used in burst-valves [34, 35] or to transport liquids back and forth by alternating between centrifugal and capillary force movement [77]. Hwang et al. [102] superpose both effects. Whereas in these approaches the rotation is purposefully generated to drive some fluid action, the present work focuses on how a given movement of the structured parts might actually hinder the lubricant flow.

All the above mentioned effects come with the disadvantage of requiring an external source of energy to drive the flow. This is not true any more for heat pipes. These closed tubes with a capillary structure inside are used for example to cool chips in electronic devices. They take advantage of the available heat as energy source. Besides capillary effects their function relies on phase transitions – therefore they have to be thoroughly closed which complicates fabrication [88].

While for this reason the present work concentrates on open microchannel flow, tube flow is its theoretical basis, cf. section 2.5.1. The century-old groundbreaking work of Lucas [134] and Washburn [204] has been confirmed in a multitude of experiments. Today, research interest in capillary tubes is concentrated more on the formation of capillary menisci especially in tubes with corners or edges. Theoretical models have been developed in [59, 156, 163] while experiments to determine static shape and flow characteristics have been done in favorable microgravity conditions, in drop towers or aboard a spacecraft [60, 124]. An advanced model, combining both bulk flow in a microchannel with corner flows has been developed and continuously refined by Weislogel and coworkers [208, 209].

The very demonstrative and probably most widespread application of capillary transport in tube-like structures is the one used by plants in their xylem, a transport tissue. Despite its obvious reliability and performance, liquid transport in plants continues to be only partly understood [56].

As channel shape is decisive both for flow resistance and magnitude of capillary force researchers typically concentrate on a characteristic geometry. Besides the standard case of a rectangular cross section, triangular [139, 167], trapezoidal [140, 182, 190] sinusoidal [182], and hexagonal [47] microchannel cross-sections have been analyzed. Axial variation in circular cross-sections was considered as early as 1921 in [96] with respect to conical shapes, while diameter variation is covered in [71, 166, 175]. Flexible walls [30], wall oscillations [6, 7] and pulsatile flows [114] have been investigated. Surface roughness in channels is researched e.g., in [91, 110]. Friction at an open channel interface is analyzed in [182, 190] while [146, 149] give attention to evaporating microflows. As each of these influences naturally adds to the complexity of flow modeling, care must be taken in the following chapters to consider only the decisive effects.

Simulation

Simulation of microscale flows is diverse. A certain overview is given e.g., in [83, 112], both pointing out the variety of approaches and the applicability of finite element, finite differences, and finite volume discretization. Comparative analysis reveals shortcomings in common commercially available CFD codes, especially with respect to surfaces and boundaries [81]. The latter are specially targeted by phase field [95, 161] and level set methods [185, 215]. Although there is broad consensus that microflows can still be treated as continua, already in the lower micrometer range the focus gets shifted towards molecular dynamics solutions and Lattice-Boltzmann methods [40, 61, 121]. For the computation of static fluid shapes, especially capillary menisci, the SURFACE EVOLVER code was written based on the energy minimization principle [23]. Building on it, Klatte et al. [116] developed an iterative procedure for channel flow including viscous friction. Literature dealing with pressure-driven channel flow which includes capillary flow as a subgroup still emphasizes the classical quasi-static Laplace or Poisson equation based approach which is adopted in the present work.

1.3. Thesis outline

Having motivated the present research and with an overview of literature given, the following chapter two sums up the existing basic theory. In order to focus on relevant concepts only, it is already somewhat selective with the respective arguments stated. Chapter three focuses on hardware and experimental design: It discusses which surface structure expectedly fulfills the task of lubricant transport towards tribocontact best and defines the microchannel as essential building block. Given engineering materials used in tribological application, section 3.2 discusses appropriate fabrication methods which can provide (more or less) satisfying structuring results. Machined test pieces are used in flow experiments described in section 3.3. Experimental results are only presented in chapter four: Based on the previous chapters, it focuses on modeling the capillary microchannel transport and compares simulation with experiments. Influences from typical application conditions e.g., temperature gradients in the surroundings of a tribocontact and movement of contact partners are considered as well as limitations from fabrication. While it becomes clear that capillary lubricant transport in microchannels towards a tribocontact can be successfully simulated, the transition into the transport structure and especially the transition out of the structure and into tribocontact offer further research questions. They are briefly addressed in chapter five before conclusions are drawn.

2. Selected topics in microfluidics and capillary flow theory

The goal of this chapter is to give an overview of microfluidics theory as far as it is relevant in the context of this work. Section 2.1 serves to focus on the right scale (e.g., arguing for a continuum instead of a molecular dynamics model), section 2.2 concentrates on the essential influencing fluid parameters and interfacial effects later modeling must build upon. Starting at the general equations of fluid motion, section 2.3 illustrates common simplifications in microfluidics and the influence of a non-inertial frame of reference. This is followed by the treatment of forces acting both in the bulk fluid and at the boundaries. Section 2.5 deals with capillary forces in more detail and completes the chapter.

2.1. Microscale characteristics

Some physical effects only become significant on a certain scale. For this reason, microfluidics research differs somewhat from classical fluid mechanics: While in the latter, inertial effects, especially turbulence, receive great attention, they hardly play a role in microfluidics. Here viscous and surface forces generally dominate.

To determine the relevance of a physical effect, it has proven useful to set it in relation with another physical effect. The capillary length for example sets capillarity and gravity in relation:

$$l_{cap} = \sqrt{\frac{\gamma}{\rho g}} \tag{2.1}$$

Here g is the gravitational acceleration, ρ is the density of a liquid e.g., oil and γ is the surface tension between the liquid and an adjacent fluid, typically air. In this case, the capillary length is around 2 mm. If a given problem has a characteristic length scale $h_0 < l_{cap}$, capillarity is strong relative to gravity i.e., l_{cap} acts as a an approximate upper bound for the exploitation of the effect. For a given length scale h_0 (e.g., the height of a microchannel) the respective dimensionless Bond number is typically preferred over capillary length. It reads:

$$\mathrm{Bo} = \frac{\rho g h_0^2}{\gamma} \tag{2.2}$$

A small Bond number, which is common in microfluidics, indicates that surface forces dominate gravity, which often allows to neglect gravitational influence altogether. Throughout this work, this assumption can be adopted based on sufficiently small Bo.

Below, capillarity is bounded once molecular effects become dominant. Used mainly for gases, a Knudsen number $\mathrm{Kn} = l_d/h_0$ can also be calculated for liquids (where l_d is then lattice spacing) but would require a characteristic length in the lower nm-range to become significant [76]. More relevant are production parameters: If microchannels are produced by laser ablation, wavelength of visible light which is in the hundreds of nanometers range is a natural lower bound, while state-of-the art micromilling tools have diameters well in the micrometer range. As the current state of the art in fabrication technology on the nanoscale is even far less developed than on the microscale, potential tribological applications are expected to emerge on the microscale. From a tribologists' view, typical film height in tribocontacts is on the order of a few microns. The Hertzian radius might be a little bigger in size while surface roughness just below the micron threshold is still quite common.

Between such bounds, the microscale is consequently the relevant scale for this work, with bounds at 1 mm and 1 µm seen only as approximate limits. A characteristic channel length l_0 can safely be allowed to be a couple of millimeters, while a characteristic channel width or height h_0 is somewhere in the mid-micrometer range.

The scale has direct consequences for physical modeling of flows: First of all, geometrical dimensions are large enough to treat the lubricating liquid as continuum. This gives rise to the classical fluid mechanics theory adapted to the microscale without the need for generally more computationally intensive molecular dynamics approaches. The basic theory is summarized in what follows.

Secondly and at least as important, it allows to decisively reduce the complexity of fluid mechanics by neglecting all physical effects which are small on the scale in question. A small Reynolds number

$$\mathrm{Re} = \frac{\rho u_0 l_0}{\eta} \tag{2.3}$$

is the basis to treat the lubricant flow as purely laminar while inertial effects such as turbulence can safely be neglected. With a characteristic speed u_0 of a few millimeters per second and the physical properties of a typical PAO lubricating oil as described in sections 2.2 and 3.3.2, $\mathrm{Re} \lesssim 1$ is expected. This characterizes lubricant liquid flow towards tribocontact as slow viscous flow or creeping flow where viscous and surface forces outplay mass accelerations and gravity.

2.2. Properties of lubricating liquids

2.2.1. Oils and greases

Oils

While solids and gases can be lubricants, e.g., graphite or air, most lubricants are liquids. The majority of lubrication fluids are oils even though any liquid can act as lubricant, for example fuel in a fuel pump. Chemically, oils are characterized by molecular chains of carbon and hydrogen atoms, irrespective of impurities. This is true for both mineral oils, found in nature, and synthetically produced hydrocarbons. The most common synthetic lubricating oils are polyalphaolefins (PAOs) [138]. Based on their frequency of use, PAOs form the basis of experiments conducted and discussed in section 3.3.

The microscale as the relevant scale of analysis throughout this work, is generally considered too large for chemical structure of the lubricating liquids to play a decisive role - in contrast to the physical properties described in the following chapters. Exceptions apply, especially if the above described base oils are enriched with critical additives. These are used to enhance specific properties like antioxidation, corrosion inhibition, antiwear behavior, or solid contaminants envelopment [138]. Mainly two groups of additives can influence capillary flow behavior significantly: First, additives which change the properties at the oils' interface to solids and to air (e.g., emulsifiers or demulsifiers, detergents, dispersants, etc.) and second viscosity modifiers which change oil viscosity significantly. This must be considered with respect to literature values and measurements of physical properties which form the basis for flow modeling.

Greases

Besides oils, greases are widely used lubricants. Greases are a compound of a solid thickener, commonly a metallic soap, and an oil which is stored inside the empty space left by the thickener (see figure 2.1).

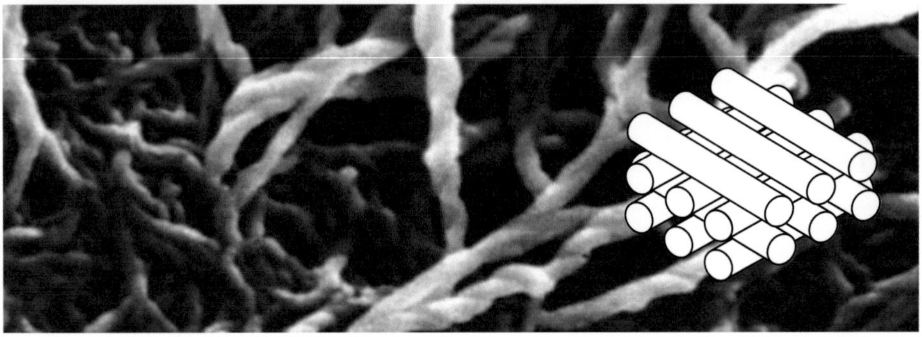

Figure 2.1.: SEM image and model of a grease: thickener structure, which serves as oil reservoir, is modeled in [5] as a stacked arrangement of fibers

If the grease is applied directly to tribocontact and is realistically expected to stay there over an extended period of time without getting pushed out of the contact by tribocontact motion, both oil and and thickener adopt lubricating tasks. If this is not the case, the oil is the sole lubricant while the thickener's purpose is to act as an oil reservoir which slowly releases the quantities of oil needed per amount of time often over a long period up to the lifespan of a product. The oil then creeps in an undirected way, drawn by surface forces towards tribocontact.

Oil bleeding of greases

The oil release or "oil bleeding" of greases is influenced by a variety of parameters: Each type of grease subjected to external pressure shows a characteristic oil release behavior as a function of time which is measured in a standardized procedure [58]. Increasing temperature facilitates oil release. The same is true if inertial forces are applied, e.g., in centrifugation [172]. In addition to measurement, Baart et al. [5] have developed an oil bleeding model for grease centrifugation based on Darcy's law, eq. (2.53), which in general governs the flow through porous media and states that the flow rate q is proportional to permeability κ_D. Modeling the grease's thickener structure as an arrangement of parallel and perpendicular fibers (as in figure 2.1) allows to cover two effects: First, the decrease of permeability as oil is released from the grease and the distances between the fibers diminish and secondly, the additional influence of tilting fibers (i.e., the change of the thickener's structural arrangement) on permeability.

As the grease's primary function is to store oil instead of transporting it over a distance, the grease itself is not in the focus of this work, except for section 5.1, where the transition of oil into a capillary is briefly discussed.

2.2.2. Physical properties of lubricating liquids

Besides the properties itself, their temperature dependence is a focus of this section. The influence of pressure, a major issue when dealing with fluid flow within tribo-contact where pressure typically reaches values in the higher MPa- or GPa-range, is negligibly small for capillary pressures, which are on or below the order of magnitude of atmospheric pressure. Viscosity for example, when modeled with the Barus ansatz, is an exponential function of pressure, with minimal differences in the low pressure range. A similar argument applies for specific heat capacity and thermal conductivity which allows to treat them as independent of pressure. If not otherwise noted all property values are given under NIST normal temperature and pressure (NTP) conditions, i.e., $T_n = 20°\text{C}$ and $p_n = 1$ atm.

Viscosity

Viscosity as a measure of the fluid's internal resistance to shearing is a decisive factor in creeping flows: When fluid layers, traveling at different speeds, are sheared, viscous friction occurs, energy is dissipated and the fluid flow slows down. Figure 2.2 shows the fluid velocities in different layers for both shear-driven (Couette) and pressure driven (Poiseuille) flows.

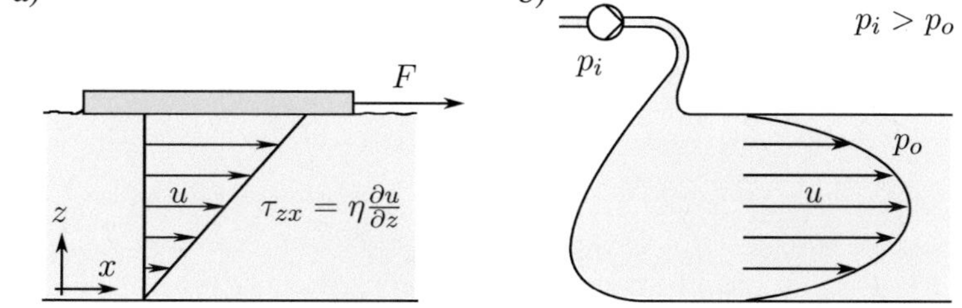

Figure 2.2.: a) Couette flow and b) Poiseuille flow with typical velocity profiles

Dynamic viscosity η is defined as the factor relating shear rate $\dot{\Upsilon}$ and shear stress τ. For the two-dimensional case this leads to

$$\tau = \eta\dot{\Upsilon} \quad \Rightarrow \quad \tau_{zx} = \eta\frac{\partial u}{\partial z} \tag{2.4}$$

Kinematic viscosity ν is the dynamic viscosity set in relation to fluid density: $\nu = \eta/\rho$. In the course of this work, the term viscosity – unless otherwise stated – always refers to dynamic viscosity. As shear stresses might occur on all sides of a fluid parcel, the equivalent to eq. (2.4) in three dimensions reads:

$$\boldsymbol{\tau} = 2\eta\boldsymbol{E} = \eta\left(\nabla\boldsymbol{u} + \nabla\boldsymbol{u}^{\intercal}\right) \tag{2.5}$$

Here, $\boldsymbol{\tau}$ and $\boldsymbol{E}$ are stress deviator tensor and strain rate tensor, respectively. For the above to be true, already two assumptions have been made: First, the fluid in question is incompressible (see next chapter for details) and secondly, it shows Newtonian behavior. A Newtonian fluid's viscosity is independent of shear rate which is not the case for a non-Newtonian fluid. Most oils are – at least in approximation – Newtonian fluids. All liquid lubricants used for experiments show Newtonian behavior, which is a basic assumption throughout this work. Greases are clearly non-Newtonian, yet are not considered in the channel transportation process.

Temperature influence on viscosity is strong: With rising temperature, viscosity diminishes significantly. The viscosity-temperature behavior is typically described by empirical models, either by Vogel-Cameron [31]

$$\eta_{VC}\left(T\right) = A_{VC}\cdot\exp\left(\frac{B_{VC}}{T + C_{VC}}\right) \tag{2.6}$$

or by Ubbelohde-Walter [194]

$$\lg\left(\lg\left(\frac{\eta_{UW}\left(T\right)}{\rho} + C_{UW}\right)\right) = K_{UW} - m_{UW}\,\lg\left(T\right) \tag{2.7}$$

The first equation is used mainly in numerical computations while the second equation forms the basis of ASTM, DIN, ISO calculation guidelines; constants A_{VC}, B_{VC}, C_{VC} and C_{UW}, K_{UW}, m_{UW} must be determined experimentally [138].

Density and incompressibility

Compressibility

$$\beta_T = -\frac{1}{V}\left(\frac{\partial V}{\partial p}\right)_T \tag{2.8}$$

of hydrocarbons is reported to lie in general below $1 \cdot 10^{-4}$ atm^{-1} [207]. With characteristic flow velocity far below the speed of sound u_s in the lubricating liquid, i.e., Mach number $\mathrm{M} = u_0/u_s \ll 0.3$, pressure influence on density can be neglected. Thermal expansion coefficients

$$\beta_p = \frac{1}{V}\left(\frac{\partial V}{\partial T}\right)_p \tag{2.9}$$

are listed for FVA reference oils (PAOs) in table 2.1 and reveal a much stronger density influence of temperature than of pressure. Density measurements for experimental fluids at various temperatures are described in section 3.3.

Table 2.1.: Compressibility, thermal conductivity at NTP conditions, and temperature influence coefficient for two FVA reference oils

		FVA 2A (PAO ISO VG 32)	FVA 3A (PAO ISO VG 100)
β_p	[1/K]	$6.91 \cdot 10^{-4}$	$6.74 \cdot 10^{-4}$
B_λ	[W/(m K^2)]	$2.96 \cdot 10^{-4}$	$2.58 \cdot 10^{-4}$
λ_n	[W/(m K)]	0.124	0.129

Figure 3.11b) shows that the temperature dependence of density is yet significantly weaker than the temperature dependence of viscosity. If variation in density can therefore be considered negligibly small, as argued in section 4.4, then the lubricating fluid is safely treated as incompressible in the sense that its density is constant both in temperature and in pressure. The incompressibility assumption simplifies the basic fluid mechanics equation framework, as the equation of mass continuity in a microchannel

$$\frac{\partial \rho}{\partial t} + \nabla \cdot (\rho \boldsymbol{u}) = 0 \tag{2.10}$$

directly reduces to

$$\nabla \cdot \boldsymbol{u} = 0 \tag{2.11}$$

Thermal conductivity and specific heat capacity

Bartel [9] specifies formulas to compute specific heat capacity at constant pressure and per unit volume instead of per unit mass $(c_p \cdot \rho)$ and thermal conductivity λ for lubricating oils. For constant pressure both simplify to a linear relationship:

$$(c_p \cdot \rho)(T) = (c_p \cdot \rho)_n \cdot (1 + B_{cp}(T - T_n)) \tag{2.12}$$

$$\lambda(T) = \lambda_n - B_\lambda (T - T_n) \tag{2.13}$$

Coefficient B_λ and normal temperature conductivity λ_n are listed in table 2.1 for the above mentioned reference oils, whereas the value of coefficient $B_{cp} = 6.5 \cdot 10^{-4}$ applies to a not further specified PAO ISO VG 150 [9]. Overall and relative relevance of property-specific temperature influence is discussed in section 4.4.2.

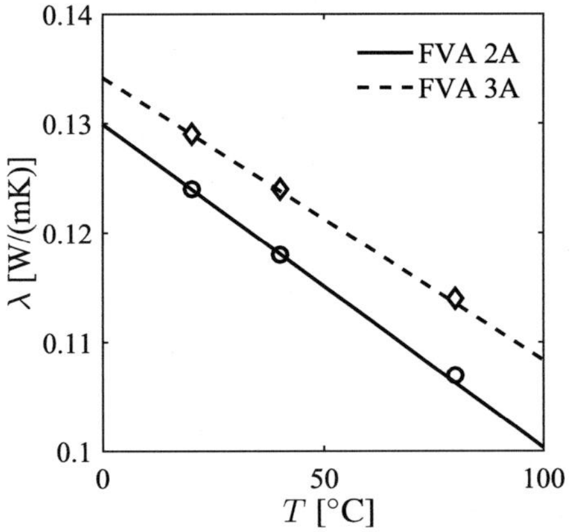

Figure 2.3.: Thermal conductivity of PAO reference oils FVA 2A (PAO ISO VG 32) and FVA 3A (PAO ISO VG 100) as a function of temperature; computation based on eq. (2.13) with measurement data from Bartel [9]

2.2.3. Interaction with solids and gases

Surface tension

If the characteristic property of stresses inside a liquid is viscosity, surface tension is the property governing the fluid's surface. From a mechanical point of view surface tension is a force per unit length whereas the thermodynamic definition considers the fact that the creation of a surface costs energy. Then, surface tension is the proportionality factor between Gibbs free energy G^I and surface area A^I and is consequently measured in energy per unit area:

$$\gamma = \left(\frac{\partial G^I}{\partial A^I}\right)_{p,T} \tag{2.14}$$

More precisely, surface tension is not a property of a liquid alone but the common property of two adjacent media which are not necessarily fluids. The more general definition of interfacial tension or interfacial energy therefore requires the definition of the phases involved i.e., $\gamma = \gamma_{la}$ as surface tension between liquid and air phases, γ_{sl} as interfacial tension between liquid and solid, and γ_{sa} between solid and air, respectively.

Contact angle

All three phases come into contact at the triple line e.g., at the edge of a water or oil drop resting on a solid surface. The equilibrium of the three interfacial tensions as material properties then defines the contact angle $\theta = \theta_{eq}$ measured within the liquid as seen in figure 2.4a). It is characteristic for the three media involved and can be determined by Young's equation:

$$\cos\theta_{eq} = \frac{\gamma_{sa} - \gamma_{sl}}{\gamma_{la}} \tag{2.15}$$

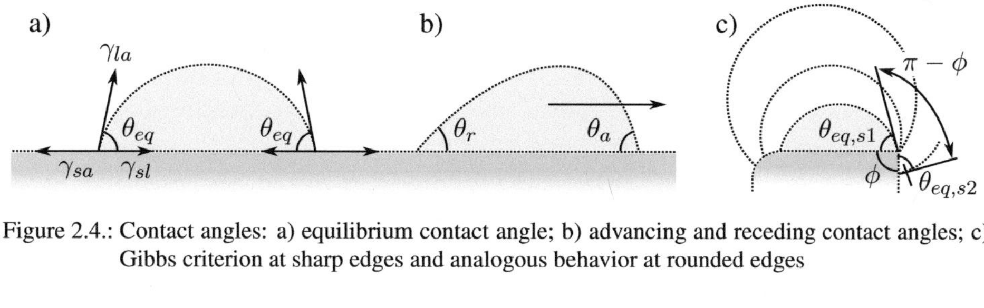

Figure 2.4.: Contact angles: a) equilibrium contact angle; b) advancing and receding contact angles; c) Gibbs criterion at sharp edges and analogous behavior at rounded edges

The contact angle can be measured comparatively easily with optical techniques. Flow dynamics alter the contact angle. If the drop moves (e.g., from left to right as in figure 2.4b), one can distinguish between advancing and receding contact angles, θ_a and θ_r respectively. Their determination is not necessarily straightforward. Favorable here is the fact that many equations, as the ones used later, only contain the cosine of the contact angle so that for small angles

$$\cos\theta_a \approx \cos\theta_r \approx \cos\theta \approx 1 \tag{2.16}$$

Thermal and pressure influence on surface tension and contact angle

As with other fluid properties, the influence of pressure on surface tension is significantly smaller than the influence of temperature. Therefore, it is a common assumption to treat surface tension as a function of temperature alone [125]. The degree of its temperature dependence is seen in figure 3.11b).

As a direct consequence, θ likewise is a function of temperature. Bernardin et al. [13] describe a low temperature regime with only slight decrease of contact angle with rising temperature, while in the high-temperature regime the negative gradient gets more pronounced.

Gibbs criterion

At a sharp edge as seen in figure 2.4c) the contact angle is undefined over a certain range, a phenomenon termed Gibbs criterion, with the range given as

$$\theta_{eq,s1} \leq \theta \leq \theta_{eq,s2} + (\pi - \phi) \tag{2.17}$$

where $\theta_{eq,s1}$ and $\theta_{eq,s2}$ are the equilibrium contact angles on the two solid surfaces which form the edge [33]. For not perfectly sharp but slightly rounded edges (as seen in figure 2.4c) on the left), the angle is never undefined but always θ_{eq}. Yet the outcome is comparable to Gibbs criterion: Even large drops, which already bend over the edge do not wet the vertical surface. Analogously, a flow can get pinned at an edge and consequently stops advancement.

Wetting

Based on the Gibbs criterion, microgeometrical surface features like pillars are investigated in order to model the wetting behavior of non-flat surfaces [67]. Surface roughness can lead to an apparent contact angle θ_{app} i.e., the one observable on a larger scale than the roughness itself, with $\theta_{app} \gtrless \theta$ to a significant degree. An overview of classical models by Wenzel and Cassie and consequences for surfaces to become more hydro-/oleophilic or hydro-/oleophobic, respectively have been summarized by Queré [162]. Also the chemical characteristics of a solid surface, especially its uppermost layer, greatly influences wetting behavior and apparent contact angle [52]. Even though tribochemistry has to deal extensively with thin solid layers chemically different from the main solid material, the effects of both surface chemistry and microgeometry on wetting and contact angle are not further discussed here: First, contact angle measurements determine the apparent contact angle which is considered more relevant than the theoretical angle and secondly, small angle influences are generally negligible, cf. eq. (2.16).

2.3. Equations of fluid motion

2.3.1. Cauchy, Navier-Stokes, and energy equations

Scaling arguments in section 2.1 already motivated the use of classical continuum theory in this context, which derives the equations of motions by applying conservation principles to mass, momentum and if necessary energy. Mass continuity already led to the incompressibility condition $\nabla \cdot \boldsymbol{u} = 0$, eq. (2.11). On its basis i.e., with constant density both in temperature and in pressure, momentum continuity yields Cauchy's momentum equation :

$$\rho \frac{\mathrm{D}\boldsymbol{u}}{\mathrm{D}t} = \rho \left(\frac{\partial \boldsymbol{u}}{\partial t} + \boldsymbol{u} \cdot \nabla \boldsymbol{u} \right) = \nabla \cdot \boldsymbol{\sigma} + \boldsymbol{f}_b = -\nabla p + \nabla \cdot \boldsymbol{\tau} + \boldsymbol{f}_b \tag{2.18}$$

$\boldsymbol{f}_b$ is the sum of body forces present. The assumption of the lubricant fluid to be Newtonian leads with eq. (2.5) to:

$$\rho \frac{D\boldsymbol{u}}{Dt} = -\nabla p + \nabla \cdot \left(\eta \left(\nabla \boldsymbol{u} + \nabla \boldsymbol{u}^{\mathsf{T}} \right) \right) + \boldsymbol{f}_b \tag{2.19}$$

In the case of constant viscosity η, further simplification is possible:

$$\rho \frac{D\boldsymbol{u}}{Dt} = -\nabla p + \eta \nabla^2 \boldsymbol{u} + \boldsymbol{f}_b \tag{2.20}$$

This is the momentum equation which together with the continuity equation (2.11) forms the Navier-Stokes equations for incompressible fluids.

Energy equation

The above equations may be complemented by the energy equation, if transport of thermal energy is of interest. The thermal contribution from viscous friction inside the fluid is generally small if temperature differences are generated also by external effects i.e., sources and sinks of energy or non-isothermal boundary conditions; neglecting viscous friction, the energy equation reads [123]:

$$\rho c_p \left(\frac{\partial T}{\partial t} + \boldsymbol{u} \cdot \nabla T \right) = \nabla \cdot (\lambda \nabla T) + Q_{th} \tag{2.21}$$

Further simplification is achieved in the stationary case, in the absence of sources and sinks of energy, and if thermal conductivity λ can be considered a constant:

$$\boldsymbol{u} \cdot \nabla T = \chi \nabla^2 T \tag{2.22}$$

Here, $\chi = \lambda / (\rho c_p)$ is the thermal diffusivity.

2.3.2. Stokes equations as a result of nondimensionalization

The universality of Cauchy and Navier-Stokes equations comes with the drawback of costly solution methods. In general the equations are further simplified by approximations, depending on the specific field of application. For slow viscous flows, Stokes equations are seen as well suited approximation of the more general equations. The simplification process is based on a nondimensionalization procedure: The basic equations are made dimensionless with respect to characteristic scales and rearranged. Then a comparison of the absolute size of different terms in the equation set is followed by the disregard of terms which are significantly smaller than others. This in the end leads to the simplified equations.

The procedure is only described here in excerpts with the complete set of nondimensionalized variables listed in the appendix. The position vector $\boldsymbol{x}$ is nondimensionalized via a characteristic length l_0 to yield the nondimensionalized position vector $\boldsymbol{x}^*$ defined as:

$$\boldsymbol{x}^* = \frac{\boldsymbol{x}}{l_0} \tag{2.23}$$

Likewise, body forces $\boldsymbol{f}_b$ are nondimensionalized by f_0. Time t is set in relation to characteristic length and velocity as $t^* = u_0 t / l_0$. As a consequence of (2.23), $\nabla^* = l_0 \nabla$. Pressure is nondimensionalized in different ways, with

$$p^* = \frac{l_0}{\eta u_0} p \tag{2.24}$$

typically chosen for Stokes flows instead of $p^* = p/(\rho u_0^2)$. With these definitions, the Navier-Stokes momentum equation (2.20) transforms to

$$\mathrm{Re}\left(\frac{\partial \boldsymbol{u}^*}{\partial t^*} + \boldsymbol{u}^* \cdot \nabla^* \boldsymbol{u}^*\right) = -\nabla^* p^* + {\nabla^*}^2 \boldsymbol{u}^* + \frac{f_0 {l_0}^2}{\eta u_0} \boldsymbol{f}_b^* \tag{2.25}$$

while continuity basically stays the same ($\nabla^* \cdot \boldsymbol{u}^* = 0$).

The simplification comes in if $\mathrm{Re} \ll 1$, which is frequently, though not always the case in microfluidics. Then the whole left side of the momentum equation is of negligible magnitude. Disregarding the respective terms and changing back to dimensional quantities yields Stokes equations:

$$\nabla p = \eta \nabla^2 \boldsymbol{u} + \boldsymbol{f}_b \tag{2.26}$$

$$\nabla \cdot \boldsymbol{u} = 0 \tag{2.27}$$

The nonlinearity due to convective acceleration has vanished. One is left with an equation system in which pressure gradients are balanced by viscous terms and – if present – body force densities.

Unidirectional Stokes equations

The long, slender geometry of a microchannel usually allows to reduce the flow field to a unidirectional one, namely downstream flow without lateral or vertical crossflows. Using a standard Cartesian coordinate system, the x-axis is chosen to be aligned with the main flow direction i.e., downstream, the height of the fluid column is oriented along the z-axis while the y-axis is in direction of the channel widths (see figure 2.5).

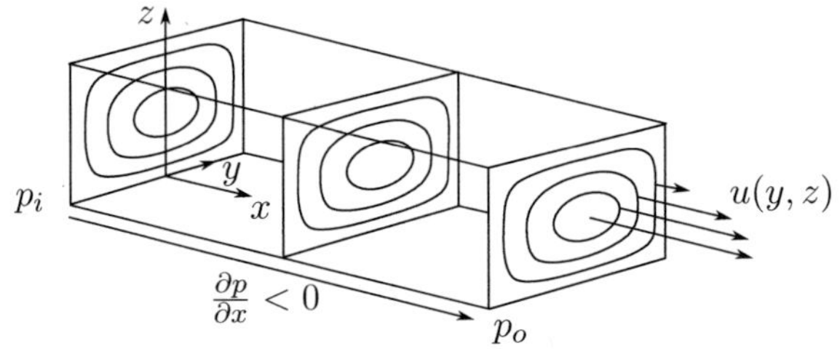

Figure 2.5.: Constant cross-section microchannel with pressure driven flow; contour lines of the velocity profile in cross-sections with maximum downstream velocity in the channel center

The unidirectional Stokes equations (in absence of body forces) then read:

$$\frac{\mathrm{d}p}{\mathrm{d}x} = \eta \left(\frac{\partial u}{\partial y} + \frac{\partial u}{\partial z} \right) \tag{2.28}$$

$$\frac{\partial u}{\partial x} = 0 \tag{2.29}$$

This equation set is commonly used as the basis for microchannel flow modeling e.g., by [27, 187, 210], even though for special applications crossflows are vital and cannot be neglected as in fluid mixing [130, 181]. The equations arc based on the assumption that a microchannel has a constant cross-section along the entire channel. Only then, the cross-sectional velocity profile as determined by eq. (2.28) is independent of downstream location and consequently serves to describe the flow in the whole channel. The non-stationary nature of a capillary flow is taken into account at a later stage and will lead to Washburn's equation (see section 2.5.1 for details).

2.3.3. Lubrication theory and Reynolds equation

In analogy to the derivation of Stokes equations, lubrication theory builds on a simplified approximation of the Navier-Stokes (or Cauchy) equations by a nondimensionalization process. Key difference is that lubrication theory allows the coordinates to have different characteristic length scales instead of the common characteristic scale l_0 used in eq. (2.23). Motivation here comes from observation of nature: There is a class of problems, where length scales typically differ significantly e.g., in the description of a thin fluid layer on a solid substrate. The characteristic height h_0 of such a layer is in general significantly smaller than the other two length scales i.e., the substrate area which is covered by the fluid layer:

$$h_0 \ll l_0 \tag{2.30}$$

is the main condition for lubrication theory. It is used extensively in thin film flow and (as the name suggests) lubrication and tribology theory yet is not limited to it. Other fields have adopted the term if the above condition is met so that the following mathematical treatment applies. It is well described in literature which derives Reynolds equation, cf. [92, 183], with only excerpts given in the following.

In the Cartesian coordinate system introduced in the previous section, $l_0 := x_0$ and $h_0 := z_0$. Without loss of generality the characteristic length scales in x- and y- directions can be defined to be equal:

$$y_{0(tf)} := x_0 \tag{2.31}$$

This still allows the actual dimensions in x and y to be of different size, yet not significantly. Pressure is nondimensionalized not via (2.24) but as $p^* = h_0^2 p/(\eta u_0 l_0)$.

With these definitions Navier-Stokes momentum equations read

$$\delta^2 \mathrm{Re}\,(\text{inertial terms}) = -\frac{\partial p^*}{\partial x^*} + \delta^2 \frac{\partial^2 u^*}{\partial x^{*2}} + \delta^2 \frac{\partial^2 u^*}{\partial y^{*2}} + \frac{\partial^2 u^*}{\partial z^{*2}} \tag{2.32}$$

$$\delta^2 \mathrm{Re}\,(\text{inertial terms}) = -\frac{\partial p^*}{\partial y^*} + \delta^2 \frac{\partial^2 v^*}{\partial x^{*2}} + \delta^2 \frac{\partial^2 v^*}{\partial y^{*2}} + \frac{\partial^2 v^*}{\partial z^{*2}} \tag{2.33}$$

$$\delta^4 \mathrm{Re}\,(\text{inertial terms}) = -\frac{\partial p^*}{\partial z^*} + \delta^4 \frac{\partial^2 w^*}{\partial x^{*2}} + \delta^4 \frac{\partial^2 w^*}{\partial y^{*2}} + \delta^2 \frac{\partial^2 w^*}{\partial z^{*2}} \tag{2.34}$$

where the parameter δ defined as

$$\delta := \frac{h_0}{l_0} \ll 1 \tag{2.35}$$

is small under the lubrication condition (2.30). Consequently, dimensionless terms which are multiplied by δ are equally small. Neglecting these terms leads to simpler equations which are still a good approximation of the original equations. The higher the power of δ, the smaller is the approximation error regarding the respective terms. As far as inertial terms are concerned, the condition for them to be neglected is not any more that Reynolds number is small (as in Stokes equation) but that the "reduced Reynolds number" [148]

$$\mathrm{Re}_r := \delta^2 \mathrm{Re} \ll 1 \tag{2.36}$$

is small – an even more likely condition. After neglecting all small terms and switching back to dimensional quantities, the resulting equation system is characterized by constant pressure across the film in z-direction:

$$\frac{\partial p}{\partial x} = \eta \frac{\partial^2 u}{\partial z^2} \quad ; \quad \frac{\partial p}{\partial y} = \eta \frac{\partial^2 v}{\partial z^2} \quad ; \quad \frac{\partial p}{\partial z} = 0 \quad ; \quad \nabla \cdot \boldsymbol{u} = 0 \tag{2.37}$$

If one allows viscosity to be non-constant, the derivation procedure has to take Cauchy's momentum equation (instead of Navier-Stokes) as a starting point as shown e.g., by Hamrock et al. [92]. With the commonly applied assumption in the computation of tribocontact film conditions, namely that viscosity can be treated constant across the film in z-direction i.e., $\eta = \eta_{avg}(x, y)$, such derivation exactly results in equation system (2.37). It is the basis for the further derivation of Reynolds equation, cf. [92, 183], which reads:

$$0 = \underbrace{\frac{\partial}{\partial x}\left(-\frac{\rho h^3}{12\eta_{avg}}\frac{\partial p}{\partial x}\right) + \frac{\partial}{\partial y}\left(-\frac{\rho h^3}{12\eta_{avg}}\frac{\partial p}{\partial y}\right)}_{\text{Poiseuille}} + \underbrace{h\frac{\partial \rho}{\partial t}}_{\text{local expansion}} +$$
$$\underbrace{\frac{\partial}{\partial x}\left(\frac{\rho h\left(u_A + u_B\right)}{2}\right) + \frac{\partial}{\partial y}\left(\frac{\rho h\left(v_A + v_B\right)}{2}\right)}_{\text{Couette}} \underbrace{+\rho\left(w_A - w_B\right) - \rho u_A \frac{\partial h}{\partial x} - \rho v_A \frac{\partial h}{\partial y}}_{\text{squeeze}} \tag{2.38}$$

Reynolds equation governs fluid flow, pressures, film height h in a tribocontact, depending on boundary conditions such as surface velocities $\boldsymbol{u}_A$, $\boldsymbol{u}_B$ of contact bodies A and B, cf. figure 1.1a). The equation combines Poiseuille and Couette flow and considers local expansion and squeeze effects. As this research is not centered on tribocontact itself but on fluid transport in the surroundings of a tribological contact, a more detailed discussion is skipped here.

2.3.4. Flow in non-inertial reference frames

So far all equations have been considered in an inertial frame of reference (I): Fluid movements have been analyzed while solid channel boundaries were considered fixed. Such an assumption cannot necessarily be upheld when a microchannel is engraved into a moving body. The cause for differences in flow behavior in inertial and non-inertial reference frames are inertial forces, sometimes called "fictitious forces" in the moving frame, namely centrifugal, Euler, Coriolis, and translational acceleration forces. They should not be confused with the inertial forces associated with the acceleration of fluid parcels within the fluid column which have been neglected in the derivation of Stokes equation (cf. section 2.3.2). The Cartesian coordinate system used so far to describe the flow, centered at the channel inlet with the x-axis pointing downstream, the y-axis in lateral direction and z upward as in figure 2.5 does not necessarily need to be changed if it is seen as a coordinate system within a non-inertial frame of reference (O), as in figure 2.6. A Cartesian system comes with the additional advantage that a constant cross-section channel does not change in lateral and vertical coordinates, whereas in cylindrical coordinates a constant channel width or height may be associated with decreasing angle the farther away from the origin. This then complicates the continuity equation.

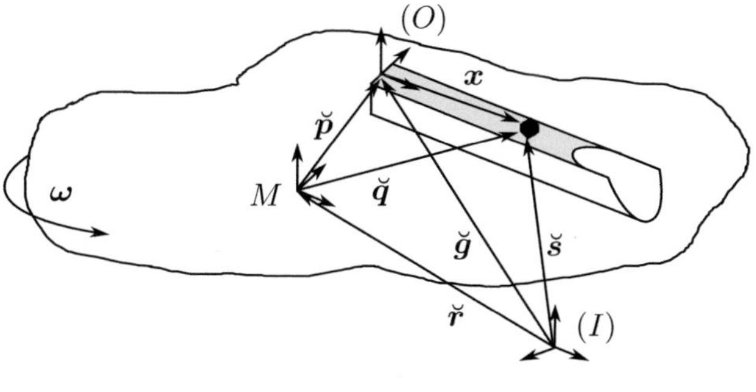

Figure 2.6.: Non-inertial channel frame of reference (O), allowed to rotate and translate relative to the inertial frame (I); M is the origin of an auxiliary coordinate system in the non-inertial frame

Non-inertial frames in microfluidics

In microfluidics literature inertial forces from frame movement receive attention in a "Lab-on-a-CD" setup: Microchannels are fabricated on a disk, which is set into rotary motion, so that the fluid is driven outward by what is termed centrifugal "pseudo" force [64] or Coriolis force is used to mix fluids in the channel. In both cases, the rotation is about an axis through the center of the disk which is fixed both in the inertial and the non-inertial frame. For channels oriented radially outward from the center of rotation, the centrifugal force density on a fluid parcel is simply determined as $\boldsymbol{f}_Z = \rho\omega^2 x \hat{\boldsymbol{e}}_x$ [63]. To account for the fluid column as a whole, the fluid parcel's position vector x is typically replaced by an averaged value for the fluid column x_{cg} [24, 64]. For channels oriented at a certain angle to the radius these terms can be updated by factoring in the appropriate (co)sine of the angle [165].

Coriolis force $\boldsymbol{f}_C = -2\rho\boldsymbol{\omega} \times \boldsymbol{u}$ is more intricate to deal with as it depends on flow velocity as a dependent variable. Two approaches can be distinguished: The first one is a two-step approach with the computation of the centrifugal force as first step which leads to the determination of downstream velocity u. On the basis of this velocity, Coriolis force is determined in a second, subsequent step [63]. The ratio of Coriolis to centrifugal force can be computed and serves as a measure of the angular velocity needed for Coriolis force to dominate [113, 169]. As such subsequent solution procedure is only an approximation of the real interdependence between forces and flows in different directions, the alternative approach is a numerical model based e.g., on the finite volume method [130, 168]. Full three-dimensional Navier-Stokes equations are used as well as models reduced to the main downstream flow direction, which are mostly similar to the unidirectional Stokes equation (2.28), updated by a centrifugal force term

$$\frac{\partial^2 u}{\partial y^2} + \frac{\partial^2 u}{\partial z^2} = \frac{1}{\eta}\left[\frac{\partial p}{\partial x} - \rho\omega^2 x\right] \tag{2.39}$$

both with and without the replacement of x by an averaged x_{cg}, cf. [64, 130, 168]. A capillary force, based on the meniscus pressure drop (cf. section 2.5) may oppose the centrifugal force, allowing to control forward and backward transport by simply controlling the rotation rate [77, 102].

General fluid motion in non-inertial frames

With the deviation from a simple rotating disk geometry, simplifications and approximations applied in such approaches are not necessarily valid any more. A more general approach is to first translate the whole governing equation set from an inertial frame to a non-inertial frame and then use appropriate simplifications and approximations. This is done for example in turbomachinery research [87, 177]. Starting point for the following analysis is the simplified Cauchy momentum equation (2.19) together with the incompressibility condition, eq. (2.11), which have been defined in an inertial frame of reference (I):

$$\rho \frac{{}^{I}\mathrm{D}}{\mathrm{D}t}\, {}_{I}\boldsymbol{u} = -\nabla p + \nabla \cdot \left(\eta \left(\nabla_{I}\boldsymbol{u} + \nabla_{I}\boldsymbol{u}^{\mathsf{T}}\right)\right) + \boldsymbol{f}_{b} \tag{2.40}$$

$$\nabla \cdot {}_{I}\boldsymbol{u} = 0 \tag{2.41}$$

Here, ${}_{I}\boldsymbol{u}$ is the absolute fluid velocity i.e., relative to the inertial frame of reference, whereas ${}_{O}\boldsymbol{u}$ is the fluid velocity relative to the non-inertial frame (O). The left superscript in ${}^{I}\mathrm{D}/\mathrm{D}t$ indicates that the material derivative is to be computed with respect to the inertial frame. Figure 2.6 depicts a situation where origin and orientation of the channel coordinate system in the moving frame (O) have been chosen as previously, namely at the channel inlet, centered laterally and fixed to the bottom of the channel. Unlike in previous sections, the channel is now part of a moving body which rotates with angular velocity $\boldsymbol{\omega}$. Additionally, the origin of the channel coordinate system translates relative to (I) as expressed in the change of the position vector ${}^{I}\mathrm{d}\breve{\boldsymbol{g}}/\mathrm{d}t$. Spatial derivatives are generally invariant with respect to the reference frame chosen: ${}^{I}\nabla = {}^{O}\nabla = \nabla$. As a direct consequence, eq. (2.41) as well as the right hand side of eq. (2.40) are valid in both reference frames. This is not true for the equation's left hand side due to the time derivative included in the material derivative. A detailed transfer from the inertial into the non-inertial frame is given in [177] and can be found in the appendix. It yields:

$$\frac{{}^{I}\mathrm{D}_{I}\boldsymbol{u}}{\mathrm{D}t} = \frac{{}^{O}\mathrm{D}_{O}\boldsymbol{u}}{\mathrm{D}t} + \boldsymbol{\omega} \times (\boldsymbol{\omega} \times \breve{\boldsymbol{q}}) + 2\boldsymbol{\omega} \times {}_{O}\boldsymbol{u} + \frac{\mathrm{d}\boldsymbol{\omega}}{\mathrm{d}t} \times \breve{\boldsymbol{q}} + \frac{{}^{I}\mathrm{d}^{2}\breve{\boldsymbol{r}}}{\mathrm{d}t^{2}} \tag{2.42}$$

In the derivation of this equation the auxiliary coordinate system with origin at M, situated in the non-inertial frame (O), was used (cf. figure 2.6). Theoretically not necessary, it comes with the advantage that instead of ${}^{I}\mathrm{d}^{2}\breve{\boldsymbol{g}}/\mathrm{d}t^{2}$ now ${}^{I}\mathrm{d}^{2}\breve{\boldsymbol{r}}/\mathrm{d}t^{2}$ must be determined. With an appropriate choice of M this may well simplify the analysis in practical application cases.

Eq. (2.42) includes various accelerations, namely centripetal $\boldsymbol{a}_{Z}$, Coriolis $\boldsymbol{a}_{C}$, Euler $\boldsymbol{a}_{E}$, and translational acceleration $\boldsymbol{a}_{T}$. Summing them up, an inertial or fictitious force density vector can be defined as

$$\boldsymbol{f}_{f} := \rho \left(\underbrace{\boldsymbol{\omega} \times (\boldsymbol{\omega} \times \breve{\boldsymbol{q}})}_{\boldsymbol{a}_{Z}} + \underbrace{2\boldsymbol{\omega} \times {}_{O}\boldsymbol{u}}_{\boldsymbol{a}_{C}} + \underbrace{\frac{\mathrm{d}\boldsymbol{\omega}}{\mathrm{d}t} \times \breve{\boldsymbol{q}}}_{\boldsymbol{a}_{E}} + \underbrace{\frac{{}^{I}\mathrm{d}^{2}\breve{\boldsymbol{r}}}{\mathrm{d}t^{2}}}_{\boldsymbol{a}_{T}} \right) \tag{2.43}$$

For given angular velocity $\boldsymbol{\omega}$, angular acceleration $\mathrm{d}\boldsymbol{\omega}/\mathrm{d}t$, and translational acceleration of M (located in the moving frame) ${}^{I}\mathrm{d}^2\breve{\boldsymbol{r}}/\mathrm{d}t^2$, the force densities based on centripetal, Euler, and translational accelerations can be treated in complete analogy to other body force densities $\boldsymbol{f}_b$ like gravity. In the case of Coriolis acceleration, which involves the relative velocity, the case is more complex and is dealt with in section 4.5. Both fictitious and body force densities can then be combined in a single force density vector $\boldsymbol{f}$:

$$\boldsymbol{f} := \boldsymbol{f}_b - \boldsymbol{f}_f \tag{2.44}$$

With this the governing momentum equation, valid in a potentially moving reference frame, becomes:

$$\rho \frac{{}^{O}\mathrm{D}}{\mathrm{D}t} {}_{O}\boldsymbol{u} = -\nabla p + \nabla \cdot \left(\eta \left(\nabla_O \boldsymbol{u} + \nabla_O \boldsymbol{u}^{\intercal}\right)\right) + \boldsymbol{f} \tag{2.45}$$

If the reference frame in consideration can be considered inertial, $\boldsymbol{f}_f = \boldsymbol{0}$. For the general validity of the equation and for the simplicity of reading, the frame index O is dropped in what follows. A major conclusion to be drawn here is that a non-inertial system can be treated just like an inertial system if a correcting term $\boldsymbol{f}_f$ is added to the momentum equation.

2.4. Acting forces in the bulk and at the boundaries

Having defined the general equation framework above, this section deals with the basics of what actually drives a microscale flow and what might hinder or even completely suppress it. Forces have to be significant and must be weighted against each other. One can distinguish between external forces acting on the fluid as body forces or at the fluid's boundaries as well as the internally generated viscous force which opposes fluid motion. Capillary force as the driver of a capillary flow typically enters the equation framework as a pressure boundary condition. This is discussed in detail in the next section 2.5.

2.4.1. Body forces and mathematical equivalents

Examples for external body forces are electric, magnetic, and gravitational forces, amongst others. Bruus [27] as well as Squires and Quake [178] provide a broad overview about possibly relevant forces in microfluidics context. In this work body forces are not in the center of attention: It has already been argued in section 2.1 that based on a small Bo number, gravity can be neglected while other body forces are of minor relevance in tribological applications. Yet as body forces simply add up in the body force vector $\boldsymbol{f}_b$, mathematical simplicity allows to carry them through the equations without effort.

Whereas inertial forces from accelerations inside the fluid column have been neglected, inertial forces which arise in non-inertial frames of reference are of special importance. Physically they are clearly different from body forces but mathematically their treatment is analogous as was seen in the previous section.

2.4.2. Flow resistance due to viscous friction

No slip boundary condition

Necessary prerequisite for viscous forces to appear is fluid motion. From the definition of viscosity, stresses are generated if a fluid is sheared as different fluid layers advance with different speeds. Decisive factor for viscous friction are boundary conditions, especially the so-called "no slip" boundary condition

$$\boldsymbol{u}_A - (\boldsymbol{u}_A \cdot \boldsymbol{n})\,\boldsymbol{n} = \boldsymbol{u}_B - (\boldsymbol{u}_B \cdot \boldsymbol{n})\,\boldsymbol{n} \tag{2.46}$$

at the boundary between two media A and B. If velocity components normal to the boundary are absent e.g., if B is an impermeable solid, the no slip condition simplifies to $\boldsymbol{u}_A = \boldsymbol{u}_B$ and can further be reduced to

$$\boldsymbol{u}_A = 0 \tag{2.47}$$

if the solid wall does not move ($\boldsymbol{u}_B = 0$). The application of the no slip boundary condition is considered to be generally valid on the macroscale, but not necessarily in small channels. Especially in nanochannels a nonzero slip length (which is of molecular dimension) and consequently a nonzero wall velocity are discussed [10]; for the microscale, the no slip condition is largely undisputed and at least an excellent approximation [125].

Based on eq. (2.4) and (2.46) uniform shear stress in a fluid between two parallel walls or within a tube leads to the typical parabolic velocity profile valid for pressure-driven (Poiseuille) flow as seen in 2.2b). With larger distance between the walls, flow velocity and throughput increase considerably underlining the influence of geometry on flow resistance.

Magnitude of the flow resistance

In the following flow resistance is discussed rather than the viscous friction force as the latter is less handy in practice. A direct measurement of viscous force or viscous stresses is – if at all possible – in most cases not straightforward. As a proxy parameter, the pressure loss between two points in the flow which is due to the energy consumed by viscous friction is commonly used, in analogy to head loss in macroscopic hydrodynamics. Pressure difference between an inlet point where $p = p_i$ and an outlet point with $p = p_o$ is related to their distance L. Assuming a linear pressure gradient allows to solve for the velocity profile in the unidirectional Stokes equation (2.28):

$$\frac{\mathrm{d}p}{\mathrm{d}x} = -\frac{\Delta p}{L} = -\frac{p_i - p_o}{L} = \eta \left(\frac{\partial^2 u}{\partial y^2} + \frac{\partial^2 u}{\partial z^2} \right) \tag{2.48}$$

This in essence states the equality of the pressure gradient and the viscous force density. As from above the solid boundary is vital for the formation of the velocity profile, the channel's cross-sectional shape (circular, rectangular, triangular, trapezoidal, etc.) greatly influences the magnitude of the flow resistance. In the special case of a laminar, axisymmetric flow of an incompressible Newtonian fluid in a tube with circular cross section of radius $b/2$, the above equation can be transferred to two dimensions in cylindrical coordinates (x, y_r) and solved analytically (as shown e.g., by Landau and Lifshitz [123]):

$$u(y_r) = \frac{\Delta p}{4\eta L} \left(\left(\frac{b}{2} \right)^2 - y_r^2 \right) \tag{2.49}$$

Volumetric flow rate q is generally determined by integrating the velocity profile over cross-sectional area A:

$$q = \int_{(A)} u \, \mathrm{d}A \tag{2.50}$$

In the case of the circular tube, combining the above equations yields Hagen-Poiseuille equation:

$$q^{(\circ)} = \frac{\pi (b/2)^4 \Delta p}{8\eta L} \tag{2.51}$$

The equation relates flow rate through a circular pipe to the pressure loss along its length by a factor which is solely determined by the tube's geometry and viscosity.

This factor is a measure for flow resistance and is comprehensibly termed hydraulic resistance [27]:

$$J_H := \frac{\Delta p}{q} \tag{2.52}$$

The proportionality between flow rate and pressure loss also applies to non-circular tubes, channels and even to volumes filled with porous media. The latter are governed by Darcy's law

$$q_D = \frac{\kappa_D A \Delta p}{\eta L} \tag{2.53}$$

which can well be interpreted as an equivalent to the Hagen-Poiseuille equation. The medium-specific permeability κ_D is typically determined experimentally. Also for channels, only for a very limited number of standard cross-sections flow rate can be analytically computed, so that the hydraulic resistance must be determined numerically or by experiments for other channel geometries (see sections 3.3 and 4.2). It may be looked up in tables or roughly approximated by $J_H \approx 2\eta L s_w^2/A^3$, cf. [27]. Instead of total channel perimeter s, the definition is based on wetted perimeter s_w i.e., the length of the solid-fluid contact line in the channel's cross-section. This also covers the possibility of a not completely closed cross-section.

Besides hydraulic resistance similar concepts have been developed in order to more easily judge viscous flow resistance in channels of arbitrary geometry and especially compare it to each other and to the base case of a circular tube. The hydraulic diameter d_H defines a virtual diameter for a channel of an arbitrary cross-sectional shape with the goal, that it is in good approximation equivalent to the actual diameter of a circular tube which produces the same pressure loss over the same length. The hydraulic diameter is defined as:

$$d_H := 4A/s_w \tag{2.54}$$

Again, tables exist for common cross-sections, cf. [41]. As an easy concept, the hydraulic diameter is widely used in literature. It yet has to be used with precaution, especially in laminar flow: Herwig [99] states a deviation of 33% from measured values even for the relatively simple case of a rectangular channel.

All deduced quantities consequently have to be handled with similar caution. This applies to the dimensionless Blasius, Moody, or Darcy friction factors

$$f^D := -\frac{2d_H}{\rho \bar{u}^2} \frac{p_o - p_i}{L} \tag{2.55}$$

as well as to Fanning's friction factor (simply one fourth of the other factors) and to the Poiseuille number $\mathrm{Po} = f^D \mathrm{Re}$. While for macroflows experimental values based inter alia on Moody's work [144] are generally considered reliable, the validity of friction factor theory in microfluidics is still an open question: After investigating more than 100 experimental studies, both results in line with theory and results which deviate highly are reported, especially with respect to laminar flow [109, 119, 179]. In conclusion, instead of a single, generally accepted reliable parameter for viscous friction or flow resistance in microfluidics there is a variety of concepts to choose from.

Frictional losses due to surface roughness

If surface roughness is larger than molecular dimensions, it is a geometrical feature of the microchannel where the no slip boundary condition applies. As surface roughness is typically significantly smaller than overall geometrical dimensions, a split-up into micro- and macrogeometry may be useful to avoid different scales in one model. In microchannels, especially when capillary effects are investigated, surface roughness is still modeled by simple geometries e.g., rectangular posts [40, 121]. Both rely on Lattice-Boltzmann models while commercial CFD codes are also applied [91, 164]. Alternatively, surface roughness is included in friction factor theory as described above [119] which reduces complexity in modeling. In laminar microchannel flow, the constricted diameter approach [110, 186] promises to be a tradeoff between results' accuracy and modeling complexity. It basically neglects the contribution of very slow flow inside surface roughness. This is done by subtracting twice the average roughness height ϵ from the nominal channel diameter b measured at the mean floor line to arrive at constricted diameter b_c (cf. figure 2.7):

$$b_c := b - 2\epsilon \tag{2.56}$$

Obviously, the error grows with increased roughness. Kandlikar et al. [110] propose a threshold of 14% relative roughness, defined as

$$\epsilon_{rel} := \epsilon / b \tag{2.57}$$

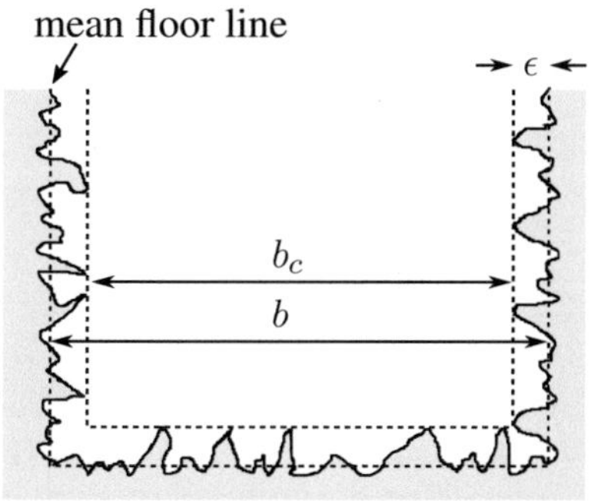

Figure 2.7.: Surface roughness in microchannels: Constricted diameter approach

2.4.3. Moving boundaries and Couette flow

The no slip boundary condition, eq. (2.46), still applies for moving walls. In contrast to static walls, moving boundaries can drive a flow even in the absence of a pressure gradient. Such Couette flow is contrasted with pressure driven Poiseuille flow in figure 2.2 which shows the typical straight line Couette velocity profile. The linear nature of Stokes equation allows superposition of Couette and Poiseuille flow. This can be seen both in Reynolds equation (2.38) and in later analysis, cf. chapter four.

2.4.4. Stress at fluid-fluid interfaces

The geometry of a solid-fluid interface is given by the solid's shape if forces are too weak to deform the solid significantly. Unlike in EHD theory, this is the case here. A fluid-fluid interface between phases A and B e.g., an oil-air boundary is defined by dynamic equilibrium conditions. The general interface stress balance reads [125]:

$$(\boldsymbol{\sigma}_A - \boldsymbol{\sigma}_B) \cdot \boldsymbol{n} = \gamma \boldsymbol{n} (\nabla \cdot \boldsymbol{n}) - \nabla_S \gamma \tag{2.58}$$

Here, $\boldsymbol{\sigma}_{A,B}$ are the respective Cauchy stress tensors, $\boldsymbol{n}$ is the interface normal, and ∇_S is the surface gradient (i.e., the gradient along an interface) which is defined as $\nabla_S = \nabla - \boldsymbol{n} (\boldsymbol{n} \cdot \nabla)$. The equation can be split up into normal and tangential directions, as shown below.

Tangential stresses and no stress boundary condition

The two tangential stress balances read

$$[(\boldsymbol{\tau}_A - \boldsymbol{\tau}_B) \cdot \boldsymbol{n}] \cdot \boldsymbol{t}_{1,2} = -(\nabla_S \gamma) \cdot \boldsymbol{t}_{1,2} \tag{2.59}$$

where $\boldsymbol{t}_1$ and $\boldsymbol{t}_2$ are two unit tangent vectors in the interface. They shall be defined to be aligned with principal directions i.e., in direction of maximum ($\boldsymbol{t}_1$) and minimum ($\boldsymbol{t}_2$) surface curvature, respectively. This implies that $\boldsymbol{t}_1 \perp \boldsymbol{t}_2$. For equal principal curvatures two arbitrary $\boldsymbol{t}_{1,2}$ may be defined; further derivation is simplified if they are analogously chosen to be perpendicular.

The tangential stress balance reveals that a surface tension gradient along the interface is prerequisite for the existence of interfacial tangential stresses. Vice versa, the absence of a gradient in surface tension is equivalent to a tangential stress free interface $[(\boldsymbol{\tau}_A - \boldsymbol{\tau}_B) \cdot \boldsymbol{n}] \cdot \boldsymbol{t}_{1,2} = 0$ in the sense that stress components are continuous across the interface.

Neglecting stresses in fluid A, for example because of a large viscosity difference $\eta_A \ll \eta_B$ as e.g., between oil and air, eq. (2.59) can be approximated with the help of eq. (2.5) by:

$$\eta_B \left[(\nabla \boldsymbol{u} + \nabla \boldsymbol{u}^{\mathsf{T}}) \cdot \boldsymbol{n}\right] \cdot \boldsymbol{t}_{1,2} = (\nabla_S \gamma) \cdot \boldsymbol{t}_{1,2} \tag{2.60}$$

In absence of a surface tension gradient $\nabla_S \gamma = 0$, the above equation is the classical "no stress" boundary condition for a fluid-fluid interface, which as Leal [125] points out is rather characterized by a zero velocity gradient. It might be more well-known in two dimensions as

$$\frac{\partial u}{\partial z} = 0 \tag{2.61}$$

Normal stresses and Young-Laplace equation

Normal stresses (due to pressure difference and fluid shearing) are balanced by surface curvature $\nabla \cdot \boldsymbol{n}$ weighted by surface tension:

$$p_B - p_A + [(\boldsymbol{\tau}_A - \boldsymbol{\tau}_B) \cdot \boldsymbol{n}] \cdot \boldsymbol{n} = \gamma \, (\nabla \cdot \boldsymbol{n}) \tag{2.62}$$

In the case of no fluid motion, this simplifies to the Young-Laplace equation:

$$p_B - p_A = \Delta p = \gamma (\nabla \cdot \boldsymbol{n}) = 2\gamma H^M = \gamma \left(\frac{1}{R_1} + \frac{1}{R_2} \right) \tag{2.63}$$

Here $2H^M$, $\nabla \cdot \boldsymbol{n}$, and $1/R_1 + 1/R_2$ all are expressions for mean curvature and R_1 and R_2 are the principal radii of curvature. To be more specific, R_1 is associated with maximum curvature, while R_2 with minimum curvature. Young-Laplace equation states that the pressure difference across a static interface is proportional to both mean curvature and interfacial tension. A flat interface, where curvature vanishes, consequently cannot support different pressures in the two adjacent fluids. On the other hand, a pressure difference in fluid statics always entails a curved interface. From eq. (2.63) curvature must be constant on the whole interface which takes the shape of a so-called "constant mean curvature surface". To determine the shape of an interface in a specific case, boundary conditions are of major importance, with details given in section 2.5.

2.4.5. Surface tension gradient and Marangoni effect

The Marangoni effect describes the phenomenon of fluid motion due to a surface tension gradient along an interface. Such a gradient in surface tension may have chemical or thermal causes: In the former, a surface-active agent or "surfactant" e.g., a detergent, changes surface tension. A nonuniform surfactant distribution along the interface leads to locally different absolute surface tension and consequently a surface tension gradient. It invokes nonzero tangential interfacial stress from eq. (2.59). The analogous is true for thermally caused surface tension gradients: Surface tension being temperature dependent only causes a nonzero resulting tangential stress if there are local gradients in temperature and consequently in surface tension along the interface. Then:

$$\nabla_S \gamma(\boldsymbol{x}) = \frac{\mathrm{d}\gamma}{\mathrm{d}T} \nabla_S T(\boldsymbol{x}) \tag{2.64}$$

If one applies this to the tangential stress balance equation of an interface (2.59) it becomes clear that a temperature gradient directly leads to discontinuous fluid stress at the interface:

$$[(\boldsymbol{\tau}_A - \boldsymbol{\tau}_B) \cdot \boldsymbol{n}] \cdot \boldsymbol{t}_{1,2} = - \left(\frac{\mathrm{d}\gamma}{\mathrm{d}T} \nabla_S T(\boldsymbol{x}) \right) \cdot \boldsymbol{t}_{1,2} := -\tau_{M_{1,2}} \tag{2.65}$$

The resulting stress $\tau_{M\,1,2}$ is termed Marangoni stress in what follows. It uniquely acts in both tangential directions at the interface without normal contribution. Consequently, the Marangoni stress cannot be balanced by pressure or other normal forces and must lead to fluid flow: It induces a shear-driven (Couette) flow in contrast to the Poiseuille flow caused by capillary action as seen in next section. As surface tension diminishes with rising temperature, thermal Marangoni flows are always directed from the warmer towards the colder region as seen in figure 2.8. Transferred to tribology, the Marangoni effect causes a lubricant flow away from the hot tribocontact to colder surrounding areas.

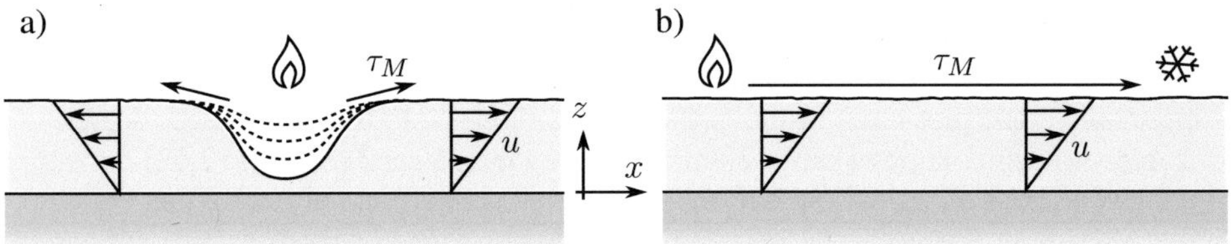

Figure 2.8.: Thermal Marangoni effect: a) local heating leads to thinning of fluid film as liquid flows away; b) temperature gradient leads to Couette flow away from hot region towards cold region

To judge the significance of the Marangoni effect, the Marangoni number is a useful tool. It reads [198]:

$$\mathrm{Ma} = \frac{\mathrm{d}\gamma}{\mathrm{d}T}\frac{l_0 \Delta T}{\eta \chi} \tag{2.66}$$

2.5. Capillary flow

2.5.1. Capillary flow in a circular tube

Capillary rise

The basic example for capillary flow is the one of capillary rise in a tube: If such a tube, open at both ends, is placed vertically into a comparatively large container filled with an appropriate liquid, then the liquid rises in the tube well above the level of the liquid surface in the container, see figure 2.9a). The force which is able to overcome gravity is capillary force and will be defined below. The capillary rise reaches its steady state at a certain height $L_{u,eq}$. Here, capillary force is exactly balanced by the weight of the fluid column inside the tube above the bulk liquid surface at $L = 0$. If the gravitational force can be neglected, the capillary force is responsible for an – at least theoretically – never stopping capillary flow, irrespective of the capillary tube or channel being in vertical, horizontal, or some other position.

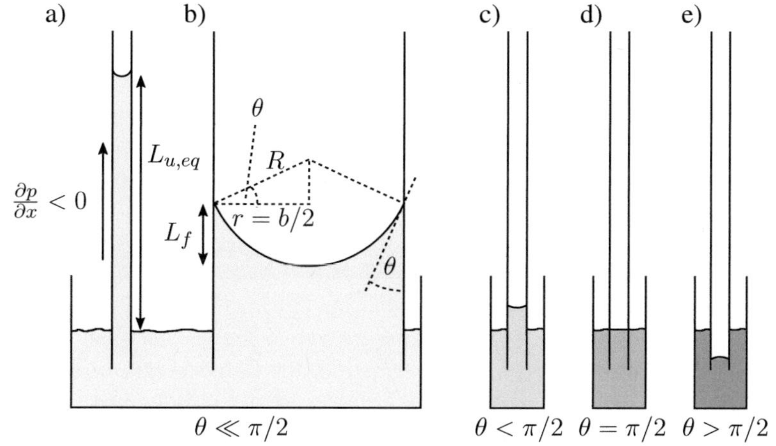

Figure 2.9.: Capillary rise in vertical circular tubes: Contact angle determines rise height

The current understanding of the origin of the capillary force has two facets: The thermodynamic approach explains capillarity-driven movement of a fluid column as the direct consequence of a system of solids and fluids trying to minimize its total Gibbs free energy, see section 2.5.5 for details. From another perspective a pressure difference at the liquid-air interface is seen as the driver of flow: At any rise height L_u, the interface is not necessarily flat but in most cases curved. This is due to the characteristic contact angle, which from Young's equation (2.15) is determined by the characteristic interfacial tensions of the materials and media involved. The contact angle over the full tube perimeter acts as a boundary condition for the liquid-air interface which is otherwise free to form and consequently obeys the general stress balance at a liquid-liquid interface, eq. (2.58). In the case of small Bo number, this results in an interface with the shape of a spherical cap, the so-called meniscus.

For a small contact angle θ, the meniscus curvature is high and diminishes gradually with rising θ until at a contact angle of $\theta = \pi/2$, the meniscus becomes flat. For contact angles $\theta > \pi/2$, the meniscus is curved again, yet adopts a convex shape instead of the concave shape for $\theta < \pi/2$, see figure 2.9 c)-e). The contact angle thus determines via meniscus curvature if capillary rise is positive or negative. The rise height is a measure for the magnitude of the capillary force.

Magnitude of the capillary force in a tube

In the static case, the capillary meniscus is governed by the normal stress balance, i.e., Young-Laplace equation (2.63). For the circular tube, the principal radii of curvature are equal ($R_1 = R_2 = R$) which simplifies the Young-Laplace equation further to yield $\Delta p^{(\circ)} = 2\gamma/R$. Here, $^{(\circ)}$ indicates that the equation is valid only in the case of a circular tube cross-section. R is related to the tube radius r by a geometrical relationship, namely $R = r/\cos\theta = b/(2\cos\theta)$ where b is the tube's diameter, see figure 2.9 b). As a consequence the meniscus pressure drop in a circular tube is:

$$\Delta p^{(\circ)} = \frac{4\gamma\cos\theta}{b} \tag{2.67}$$

As the upper part of the tube is open, pressure above the meniscus is just ambient pressure p_{am}. Then from Young-Laplace equation, the liquid pressure just below the meniscus must be $p_{am} - \Delta p$. This is an underpressure not only compared to the surrounding air but also compared to the fluid in the container the tube is placed in. At its surface, the fluid in the container has a pressure level equal to air pressure. As the tube's lower end is open, the same pressure $p = p_{am}$ also applies inside the tube at the same height. Consequently, there is a remarkable pressure difference inside the fluid-filled tube between the meniscus location and the location where the tube penetrates the surface of the fluid in the container. It is this pressure difference which drives the capillary flow upward and which is the underlying physical principle of capillary action.

From a modeling perspective, the meniscus is treated as a pressure boundary condition to the fluid-filled domain. As the meniscus' vertical extension is generally small compared to the rise height $L_f \ll L_u$, it is generally neglected, at least in the basic model presented here. The pressure difference at the meniscus then acts normal to the tube's circular cross-section A, so that the driving force for the fluid flow i.e., capillary force, becomes:

$$F_{cap} = \Delta p \cdot A \tag{2.68}$$

Capillary flow, in the absence of other forces, is purely pressure-driven i.e., a Poiseuille flow.

Capillary flow dynamics: from Hagen-Poiseuille to Lucas-Washburn equation

The filling of a capillary tube or channel is a continuous process in various stages. For all stages, the capillary force is the driving force and its magnitude stays the same if cross-sectional geometry of the channel – as assumed – stays constant. At the very beginning when the fluid is still at rest or shortly thereafter i.e., as long as fluid column length is still very small $L_u \ll l_0$, the capillary force is responsible for huge accelerations because of minimal fluid volume and mass. In short, inertial effects dominate. This phase may be neglected as from dimensional analysis above viscous forces very soon become dominant compared to inertial forces (expressed in low Reynolds or reduced Reynolds numbers) and as such the relevant counterpart of capillary force. As the total viscous force increases over time with growing fluid column length $L_u(t)$ but the capillary force stays the same, the fluid flow slows down gradually.

In a circular tube, capillary force is expressed by eq. (2.68) independent of time t. The volumetric flow rate of the tube $q(t)$ is governed by Hagen-Poiseuille equation (2.51), now in a time dependent form with $L_u(t)$. It can also be stated as the time rate of change of liquid volume in the tube which is equal to cross-sectional area A multiplied by the cross-sectional average velocity $\bar{u}$:

$$q(t) = \frac{\mathrm{d}V(t)}{\mathrm{d}t} = A\frac{\mathrm{d}L_u(t)}{\mathrm{d}t} = A\bar{u}(t) \tag{2.69}$$

This equation applies to all ducts or tubes with constant cross-section. Combining it with the above equations (2.67) and (2.51) leads to an ODE for flow dynamics:

$$\frac{\mathrm{d}L_u(t)}{\mathrm{d}t} = \frac{q(t)^{(\circ)}}{A^{(\circ)}} = \frac{b^{(\circ)}\gamma\cos\theta}{8\eta}\frac{1}{L_u(t)} := \frac{W^{(\circ)}}{2L_u(t)} \tag{2.70}$$

This equation was first stated in similar form in the groundbreaking articles of Lucas [134] and Washburn [204]. With initial condition $L_u(t=0) = 0$ it can be integrated to result in Lucas-Washburn or just Washburn equation:

$$L_u(t) = \sqrt{\frac{b^{(\circ)}\gamma\cos\theta}{4\eta}t} = \sqrt{W^{(\circ)}}\sqrt{t} \tag{2.71}$$

This equation is the cornerstone of capillary dynamics. Derived and experimentally confirmed for a circular tube, researchers continue to report that the basic time law of a square root function

$$L_u(t) \propto \sqrt{t} \tag{2.72}$$

turns out to be still valid in a variety of experimental variations e.g., with other channel cross-sections and is termed "Washburn behavior" in what follows. W is a factor combining multiple parameters which essentially governs the speed of fluid advancement.

Note that this approach is quasi-stationary: The time function is not derived from inertial dynamics of decelerating fluid mass but from an integration of stationary flows with balance of capillary and viscous forces at every point in time. This is justified by the neglection of inertial forces based on Reynolds or reduced Reynolds numbers. For an analysis including the short-time behavior at the initial stage of a capillary flow which must consider inertia, Bosanquet's solution applies [22]. For completeness, models which additionally consider the influence of gravity in capillary rise, including fluctuation around equilibrium height, are described by Ouali et al. [151] or Zhmud et al. [220].

2.5.2. Capillary force determination from meniscus pressure drop

Meniscus shape in arbitrary channel geometries

So far, all equations have applied to tubes with circular cross-section. The deviation from such basic geometry has two implications: First, viscous friction and consequently flow velocity and flow rate all depend on the cross-sectional shape of the channel. This is true for any Poiseuille flow, whether capillarity-driven or not (cf. section 2.4.2). The capillary flow's second dependence on channel geometry is due to the driving meniscus: Magnitude of meniscus pressure drop and consequently capillary force are generally (not only in the special case of a circular tube) determined by eq. (2.63) and (2.68), at least if static conditions can be assumed, as well as the contact angle boundary condition, if the channel cross-section is completely bounded by a solid wall. The contact angle itself can be measured or computed from material properties in the straightforward way outlined in section 2.2.3. The determination of the meniscus shape, given the contact angle boundary condition, is not trivial for an arbitrary cross-sectional channel geometry even though it is known that the meniscus must have constant curvature (cf. section 2.4.4). Defining the meniscus surface explicitly as a function of lateral and vertical channel coordinates $U(y, z)$, the general problem can be stated as follows [73, 74]:

$$2H^M = \nabla \cot \Gamma U \tag{2.73}$$

where

$$\Gamma U = \frac{1}{\sqrt{1 + |\nabla U|^2}} \nabla U \tag{2.74}$$

The contact angle boundary condition reads

$$\boldsymbol{n} \Gamma U = \cos\theta \tag{2.75}$$

where $\boldsymbol{n}$ is the interface exterior normal.

Solution to this problem set is sought in discrete differential geometry by energy minimizing algorithms. In microfluidics, Brakke [23] follows this approach with his SURFACE EVOLVER code which uses a variety of energies including surface tension and pressure energies and iterates towards the solution via a gradient descent method. Figure 2.10 shows the meniscus shape as interface between liquid and air phases computed with SURFACE EVOLVER in a closed channel segment with rectangular cross-section. Besides the main or bulk meniscus, liquid "fingers" form in the channel edges which will be discussed more in detail in section 2.5.3.

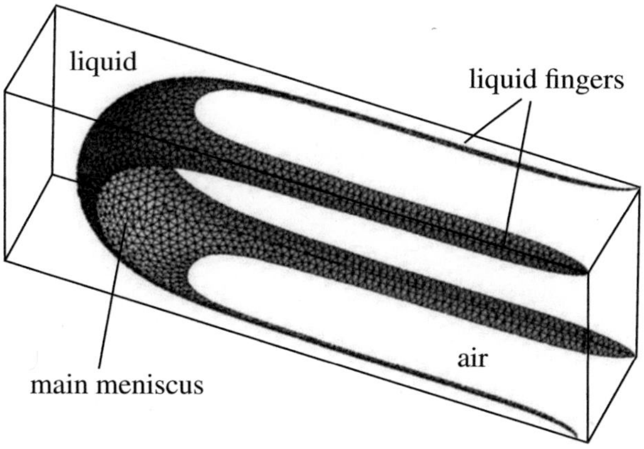

Figure 2.10.: Meniscus in a closed rectangular channel segment: Liquid-air interface with constant mean curvature computed with SURFACE EVOLVER [23] for a contact angle of 5°

Such meniscus is static as the fingers have reached the channel's end. It is not necessarily similar to an advancing meniscus where advancing contact angles differ from θ_{eq} and normal as well as tangential stresses at the interface have to be considered based on the general equations (2.59) and (2.62). This is discussed in more detail in section 4.3. Microfluidics literature typically forgoes the exact determination of the meniscus shape and instead concentrates on ways to determine the magnitude of meniscus pressure drop which can then be used – in fine analogy to capillary flow in a circular tube – as pressure boundary condition for the flow domain. This is straightforward in special cases:

Meniscus pressure drop in simple geometries

The base case of a capillary with circular cross-section, where the meniscus has spherical cap shape, has already been discussed above. The pressure drop is based on Young-Laplace equation exactly computed by eq. (2.67). An elliptical cross-section, though not a typical case in microfluidics literature, could be treated analogously yielding $\Delta p^{(ell)} = 2\gamma \cos\theta \, (1/b + 1/h)$ where $b/2$ and $h/2$ are semi-major and semi-minor axes respectively.

To emphasize the influence of a channel's cross-sectional shape in contrast to the influence of channel size and physical properties Sammarco and Burns [173] use a general form of Young-Laplace equation with geometrical constant G:

$$\Delta p = \underbrace{G}_{\text{shape}} \cdot \underbrace{\frac{1}{b}}_{\text{size}} \cdot \underbrace{\gamma \cos\theta}_{\text{phys. properties}} \tag{2.76}$$

Like this, the pressure generation of circular closed channels can be directly compared to other cross-sections – in lose analogy to hydraulic diameter or hydraulic resistance concepts applied in the context of viscous friction. Table 2.2 lists respective values of G for various channel cross-sections. Noteworthy is the channel of unlimited height i.e., two plates with distance b. Here, the second principal curvature vanishes, $R_2 \to \infty$, so that with $G = 2$ it produces only half the capillary pressure compared to a circular tube with diameter equal to the plates' distance. Expressing the relationship between channel height and channel width by the parameter aspect ratio $\Lambda := h/b$ leads to the formula given below for elliptical channels which includes the circular channel as a special case where $b = h$. The aspect ratio as a shape characteristic thus allows to measure the channel size by a single variable (either h or b), cf. eq. (2.76).

Table 2.2.: Comparison of the magnitude of the front meniscus pressure drop in channels with different cross-sectional shapes (all closed or of unlimited height) by help of the geometrical constant G, see also table 4.3

shape	circular ○	elliptical 0	two plates | |	closed square □
based on	Young-Laplace eq. (2.67)	Young-Laplace see text above	Young-Laplace see text above	Lenormand et al. eq. (2.77)
G	4	$2\left(1+\Lambda^{-1}\right)$	2	≈ 3.77

Unlike in the above cases the meniscus pressure drop inside a rectangular channel cross-section, cf. figure 2.10, cannot be expressed by simple yet exact theoretical formula any more which led to the development of approximations: Birdi [19] uses regression analysis to determine constants based on experiments with square tubes of side length between 300 μm and 1 mm. Other approaches approximate the rectangle with sides b and h by its inscribed ellipse and use the latter's exact formula as given in table 2.2, although knowing that an error is involved [103, 173]. Lenormand et al. [128] introduce an additional error minimizing factor $K_{Len}^{(\square)}$ (see figure 2.11) which also includes aspect ratio Λ. Their formula reads:

$$G_{Len}^{(\square)} = 2\left(1+\Lambda^{-1}\right) K_{Len}^{(\square)} \tag{2.77}$$

where

$$K_{Len}^{(\square)} = \frac{\Lambda(4-\pi)}{2(\Lambda+1)\left\{(\Lambda+1) - \left[(\Lambda+1)^2 - \Lambda(4-\pi)\right]^{1/2}\right\}} \tag{2.78}$$

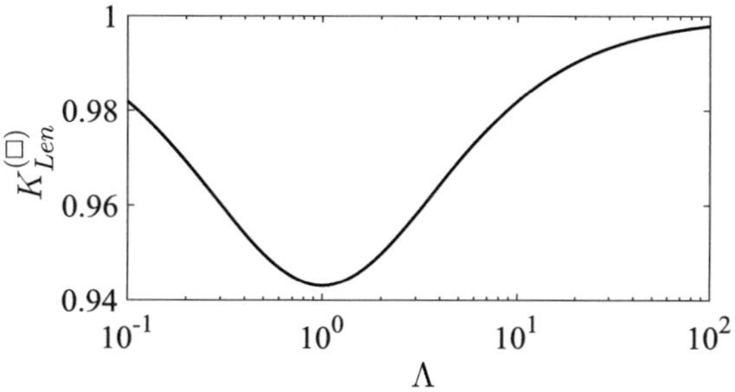

Figure 2.11.: Aspect ratio adjustment factor $K_{Len}^{(\square)}$ from [128], computed with eq. (2.78)

Originally restricted to perfect wettability, $\theta = 0$, the formula has been adapted for use in eq. (2.76) under the small angle assumption, eq. (2.16).

2.5.3. Differences in driving mechanism: Bulk and corner flow

So far it was assumed that the capillary flow is exclusively driven by a meniscus which forms the liquid-air interface at the front of the liquid column. In the following, this mechanism is called bulk (meniscus) flow, adopting Weislogel's [208] terminology. An alternative mechanism, termed corner flow, is a capillary flow characterized by a meniscus located at the side of the liquid column.

Figure 2.12 contrasts both mechanisms as they are understood in the context of this work: In bulk meniscus flow, the interface's center(s) of curvature (at least approximately) lie in main flow direction while in corner flow, the surface normals are perpendicular to main flow (again at least in approximation). As a consequence, the driving mechanism in corner flow is somewhat different from bulk meniscus flow. As seen in figure 2.12b) the main contribution to mean curvature lies in the cross-sectional plane whereas the curvature contribution along the groove is small, especially if the groove is relatively long compared to its width and height. Consequently, in good approximation the second principal curvature vanishes ($R_2 \to \infty$) whereas R_1 comes to lie in the cross sectional plane. R_1 itself is not constant along flow direction but diminishes gradually, so $R_1(X_1) > R_1(X_2) > R_1(X_3)$. Then the corresponding pressure drop between liquid and surrounding air must rise as from eq. (2.63) $\Delta p = \gamma / R_1(x)$ [59]. This leads to a negative pressure gradient in flow direction, which then – perfectly in line with bulk meniscus flow – drives the capillary flow forward.

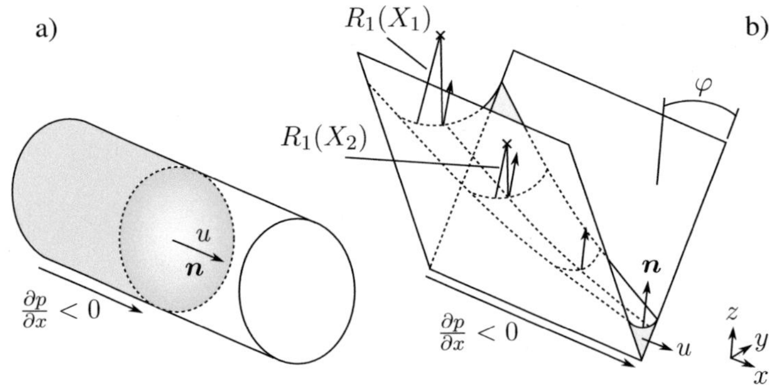

Figure 2.12.: Bulk and corner flow

Obviously, such reasoning conflicts with the notion of the meniscus being a constant mean curvature surface. As a direct consequence, there must be fluid flow (see section 2.4.4). Static equilibrium is theoretically never reached as the liquid column lengthens infinitely. This is true provided the fluid wets the corner at all which is determined by the Concus-Finn criterion [43, 54, 73]:

$$\varphi + \theta < \frac{\pi}{2} \tag{2.79}$$

Here, φ is half edge angle. In practice, such argument breaks down at the latest when molecular forces come into play, most probably earlier caused by geometrical deviation from sufficient surface roughness or theoretical edge geometry. Yet geometrical imperfectness of the corner does not necessarily prevent corner flows with details about corner "roundedness" given in [59, 163].

In general, corner flows are present in two different scenarios: First, if the channel itself has v-shaped or triangular cross-section, flow in this channel can be entirely explained by the corner flow mechanism without the need for a front meniscus [167], cf. figure 2.12b). Secondly, if the channel in consideration has pronounced edges or corners: Then, as seen in figure 2.10, liquid fingers form which advance as corner flows. As the Concus-Finn criterion applies, the fingers in this figure are bounded and the interface has constant curvature only for the fact that the channel segment has finite length. Liquid fingers and associated corner flow in this case complement bulk meniscus flow.

The simultaneous appearance of both flow mechanisms has been investigated in closed channels with multiple interior edges with the result that both flows advance similarly, namely proportional to the square root of time thus showing Washburn-like behavior [208]. This is especially remarkable as the nature of both flows as described above is quite different.

2.5.4. Meniscus configurations along the fluid column

In contrast to closed channels, which in their cross-sections are completely bounded by solid walls, open channels feature a fluid-fluid interface between liquid in the channel and typically air above it, so that a meniscus can form. Experimental evidence for such upper menisci is given by microflow analysis with the help of micro-PIV [218] with special attention to meniscus shape measurement [214] or alternatively by upper meniscus visualization via the projection of a laser onto the channel [75]. Condition for the meniscus is – from Young-Laplace equation (2.63) – a difference between liquid pressure p in the channel and outside air pressure p_{am}. Additionally, the degree of filling or saturation of the channel has to be taken into account: Saturation may vary along the channel and is defined as the fraction of the channel's cross-section which is filled with liquid, see figure 2.13a):

$$S(x) := \frac{A_S(x)}{A} \tag{2.80}$$

For a channel of constant cross-section in downstream direction, A is per definition not a function of x. The various configurations which might develop depending on saturation are discussed in [146, 187].

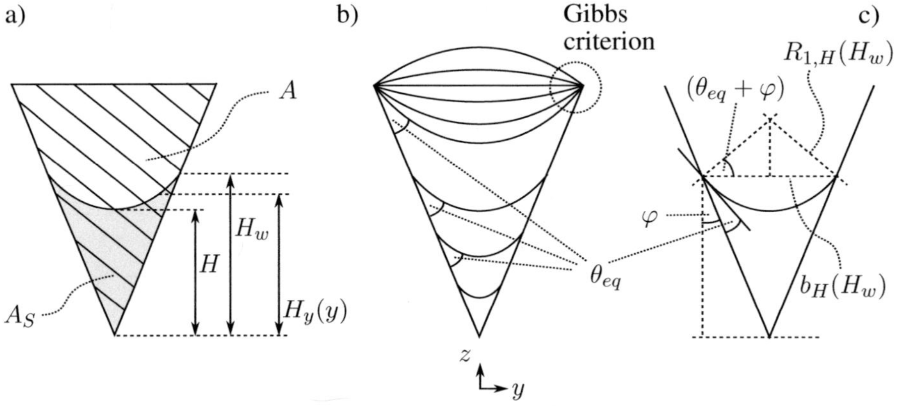

Figure 2.13.: Partially filled channel cross-section

If the meniscus is attached to the top edges of the channel, meniscus curvature can vary as depicted in figure 2.13b). This is a consequence of Gibbs criterion (for ideally sharp edges) or the contact angle boundary condition at a small edge radius, cf. section 2.2.3. In such case the meniscus radius of curvature R_1 is exclusively governed by the pressure drop across the upper liquid-air interface Δp_u. For long and slender microchannels where $z_0 \ll x_0$ it is commonly assumed that the interface is hardly curved in channel downstream direction

$$\frac{\partial^2 H}{\partial x^2} \approx 0 \tag{2.81}$$

so that its contribution to mean curvature is negligible and $R_2 \to \infty$. In perfect analogy to the argumentation used in the previous section regarding corner flows, $R_1(x)$ as a function of x is assumed to lie in the respective cross-sectional plane at x. Then

Young-Laplace equation reduces the problem to two dimensions in channel cross-section with

$$R_1(x) = \frac{\gamma}{\Delta p_u(x)} \tag{2.82}$$

If the saturation falls below a threshold level, the meniscus detaches from the upper channel edges and is then exactly determined by the contact angle θ_{eq}. The slope of the channel wall, depending on φ, must be considered as well as the diminishing channel width $b_H(H_w)$ with declining saturation [167]. From figure 2.13c), the meniscus curvature in such case is

$$R_1(H_w) = \frac{b_H(H_w)}{2\cos(\theta_{eq} + \varphi)} \tag{2.83}$$

2.5.5. Thermodynamic approach for capillary force determination by free energy minimization

So far the capillary force has been associated with the pressure drop across a meniscus. An alternative way to compute capillary force is the thermodynamic approach outlined in the following. Its great advantage is that it does not require knowledge about the exact meniscus shape.

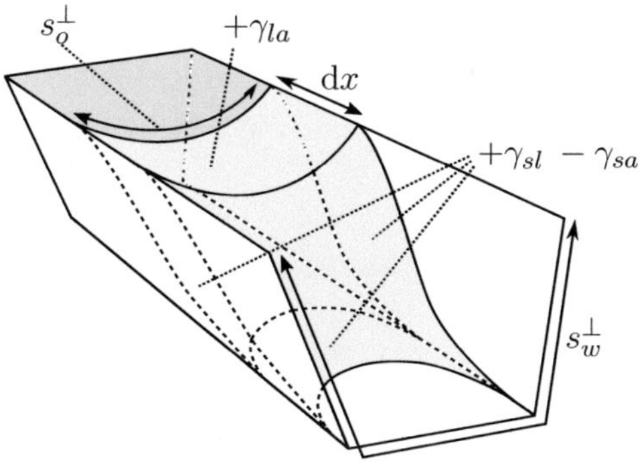

Figure 2.14.: Capillary force determination by free energy minimization approach

From the nature of surface tension it is known that the creation of a new surface A_k^I costs energy, cf. eq. (2.14). If a capillary flow advances in a channel, constantly new interfaces are created while others are dissolved. Figure 2.14 compares two states of the fluid column, before and after it has advanced by a distance $\mathrm{d}x$. On the upside of the fluid, the advancement created the new interface of area $s_o^{\perp}\mathrm{d}x$, separating air from liquid. On the left, the right and the lower channel walls, the formerly solid-air interface gets replaced by a solid-liquid interface of area $s_w^{\perp}\mathrm{d}x$. $s_o^{\perp}$ and $s_w^{\perp}$ are the projections of the liquid-air boundary line and the liquid-solid boundary line (the so-called wetted perimeter) onto the channel's cross-sectional plane. The underlying assumption that the shape of the front meniscus does not change is considered reasonable for small advancements $\mathrm{d}x$. Then a balance of the Gibbs free energies for the three interfaces $k = \{sa, sl, la\}$ yields [14, 151]:

$$\mathrm{d}G^I = \sum_{k=1}^{3} \mathrm{d}G_k^I = \sum_{k=1}^{3} \gamma_k \mathrm{d}A_k^I = \mathrm{d}x \left[\left(\gamma_{sl} - \gamma_{sa}\right) s_w^{\perp} + \gamma_{la} s_o^{\perp} \right] \tag{2.84}$$

As the capillary force is the only force doing work in this context, it must be responsible for the new interface creation:

$$F_{cap} = -\frac{\mathrm{d}G^I}{\mathrm{d}x} = \gamma_{la} \left(\cos\theta s_w^{\perp} - s_o^{\perp} \right) \tag{2.85}$$

Here, Young's equation (2.15) has been applied. The above result is only valid under the condition that $(s_o^{\perp}/s_w^{\perp}) < \cos\theta$, which for a common channel is typically unproblematic, especially if the contact angle is small.

3. Surface design, fabrication, and experimental design

This chapter deals with the preparation and execution of experiments. It starts with a discussion about surface structure design: How should a favorable surface geometry look like so that it performs well in supplying a tribocontact with lubricant? Section 3.2 then discusses the question how such a structure can be fabricated with given restrictions, e.g., when using materials typically found in tribological environments. Test pieces are produced for use in experiments, illustrated in section 3.3. Experimental results are only presented in the next chapter, where they can be compared directly with results from simulation.

3.1. Capillary transport structure design

Ultimate goal is to supply a tribocontact with liquid lubricants out of a distant reservoir. The capillary force has been identified as a possible driver of lubricant flow, even against counteracting forces which result e.g., from the Marangoni effect. As has been seen in the previous chapter, the geometry is vital in the creation of a capillary force. Consequently, the choice or design of a well suited if not best suited transportation geometry needs explicit attention.

A geometry is well suited if it excels in its task of supplying tribocontact all the while coping with the restrictions from a typical tribological environment and from production (see next section). As central criterion to judge the transportation capability of a capillary geometry the flow rate q will be used in what follows: The structure must supply the amount of lubricant the tribocontact needs in a given amount of time. In some cases, the capillary force may be of higher importance than the flow rate. This is discussed in more detail in the next chapter. Before dealing with surface structuring itself, a brief discussion of possibly applicable alternative methods is given:

Transportation methods of liquids on solid surfaces

Both a thermal gradient and surface chemistry designed e.g., by applying appropriate coatings, may allow to move liquids such as water or oil drops on solid surfaces, cf. literature section 1.2 for examples. Due to the specific conditions in the vicinity of a tribocontact both methods are not pursued further here: First, the thermal field must be considered as given. Secondly, geometrical features are expected to survive the harsh environmental conditions with for example surface contamination and wear better than chemical ones, although this is not necessarily the case. If the solid surface is not flat but curved, it is to a limited degree able to transport fluids by surface tension depending on local curvature [197]. As the macrogeometrical shape of tribocontact's contact partners is often subject to narrow design limitations, concentration on microgeometry is considered a more flexible approach. Microgeometrical features limited to a single layer at the surface of a part must be distinguished from features which reach in multiple layers into the depth of the part. The latter is covered by the theory of porous media. Transport in porous media is undirected (see below) and can be comparatively slow. A tribological example are sintered bearings. Their primary task is oil storage while the transportation to the tribocontact is of secondary priority. Additionally, structural integrity of porous media is generally lower than in solid bodies with microgeometrical features just at their surface.

Geometry influence on wettability

In absence of microgeometrical features i.e., on an ideally smooth, flat surface, the flow rate is zero: A drop of oil put on such surface flattens under gravity influence till it reaches its equilibrium position determined by the equilibrium contact angle as boundary condition. Apart from such stretching out in all directions, the drop does not move in a certain direction. The same is true for isotropic surfaces, for example an arbitrarily rough surface or a surface with intentionally created surface features. Both may alter the apparent contact angle and consequently the wettability of a surface. Changing surface design towards oleophilicity e.g., based on Cassie and Wenzel models, enhances spreading behavior of an oil drop, yet still does not create directional transport: Consequently, efficiency is low as only a relatively small fraction of available lubricant is transported to the desired destination (even though this may be acceptable and sufficient in certain applications).

Directed transport

Figure 3.1a) shows a typical surface structuring pattern: Pillars or posts of circular, square, octagonal or any other shape which are slightly higher than the surrounding surface and which allow menisci to form between them, so that the liquid is drawn forward by capillary force [45, 108, 171]. In essence, the transport is still not directed, which is healed in practice by not structuring the whole surface but only a certain area between source and destination of the liquid. Figure 3.1a) falls short in explaining the details of three dimensional capillary transport in such structures. It yet shows that different local transport directions add up to a general preferential transport direction (from left to right). Anisotropic surfaces indeed allow for directed transport: In contrast to single droplet movement [44, 211], Jokinen et al. [107] are able to generate an anisotropic flow by modifying the shape of the above mentioned pillars. As with the circular pillars in 3.1a), meniscus curvatures change depending on advancement position which leads – as a consequence of Young-Laplace equation (2.63) to different magnitudes of capillary pressure (illustrated by arrow length).

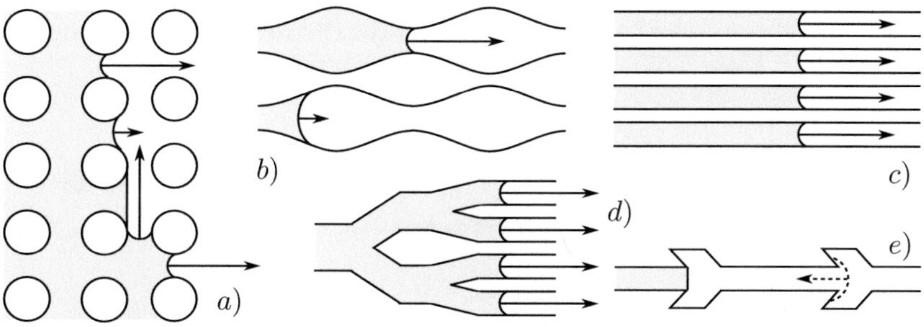

Figure 3.1.: Surface structures to transport liquids: a) pillars, b) converging-diverging channels, c) constant cross-section channels, d) tree-like structure, e) flow stopping geometry

Advantages of microchannels as depicted in figure 3.1c) are firstly, that they keep the same small and therefore strong meniscus during the whole transportation process and secondly, that capillary force is always directed downstream, towards transportation destination. Thirdly, microchannels can be densely packed, so that the overall flow rate over a given surface area is high. The amount and dimensions of the individual channels in the microchannel array depicted is basically a tradeoff between viscous resistance and magnitude of capillary force, details are given in chapter four.

Deviations from the simple constant cross-section channel are not considered advantageous: The diverging-converging capillaries depicted in figure 3.1b) need more surface space and capillary pressure varies due to the changing meniscus curvature. Only at its maximum value (at the channels' narrowest points) the meniscus pressure drop is not inferior to the value of the channels in c). Lubricant velocities in recent experiments by Hermens et al. [98] in non-constant cross-sectional channels designed to mimic lizard skin are smaller by about one order of magnitude compared to constant cross-section channel experiments described in the following section, yet are inter alia based on a different liquid so that absolute comparison is difficult. In analogy to macroscopic hydrodynamics, channel bends and turns may add to flow resistance, with more details given in section 4.2.3. Channel crossings are neither beneficial [128]. To what degree the tree-like structure depicted in figure 3.1d) which might be a way to approach the above mentioned tradeoff between capillary and viscous forces is hindered by y-shaped connections might be an interesting research question remaining unanswered in this work. Not to neglect is the possibility of stopping a capillary flow completely by appropriate geometrical design [70, 137, 222] as seen in figure 3.1e). Here, the meniscus curvature switches from concave to flat and – if additional outside pressure is to be supported – potentially even to convex. Purposefully designed, such geometry can be used as a diodic flow element which allows flow in one direction and stops flow in the opposite direction. Less likely but still possible is that geometrical imperfections – maybe from fabrication as seen in the next section – lead to a similar, then undesired effect. In such case a multitude of channels provides additional security which only adds to the advantages of a microchannel array. Therefore, such an array and the microchannel as its basic building block form the basis for further analysis.

3.2. Materials, microchannel fabrication, and geometry measurement

At least two reasons suggest that the microchannel's cross-section should preferably be small: Firstly, capillary pressure is higher the smaller channel cross-sectional dimensions. Secondly, mechanical operating conditions in tribological contacts urge to weaken structural integrity of parts to the least possible extent. Downsizing is limited mainly by production possibilities and desired accuracy. Both at least partially conflict with materials used in tribological environments:

Material restrictions as limiting factor for photolithographic and etching processes

Typical materials in tribology are metals, ceramics or engineering plastics. With respect to a certain application, materials are carefully chosen in order to fulfill an array of requirements: They must withstand high mechanical loads in a tribocontact, be sufficiently elastic, show limited wear, etc. Consequently, production technology must comply with given materials. In the following, fabrication methods are judged by their ability to successfully machine a variety of these materials. Furthermore, the technologies evaluated under laboratory conditions should at least show the potential for later application in volume production.

The former is the decisive disadvantage of maybe the most important class of microchannel fabrication technologies, namely combined photolithographic and etching processes. Here, various layers of different materials are deposited on a substrate, typically a wafer. An optical mask determines the light exposure pattern of a photo resist thereby defining the areas where lower lying layers are removed in a subsequent chemical etching process. Especially process derivatives like DRIE allow the fabrication of microchannels with very exact, predominantly rectangular cross-sectional geometry (see figure 3.2, small picture) all the while allowing for high aspect ratios in the range of $\Lambda \approx 100$ [152]. Yet, the technology is limited to a narrow class of appropriate materials such as metalloids, in particular silicon but also some polymers like PEEK or PMMA [193]. Though the LIGA process variant allows to produce small parts out of metal, it is galvanically deposited on a substrate, which is seen critical in tribological applications both for manufacturing reasons and fulfillment of mechanical requirements.

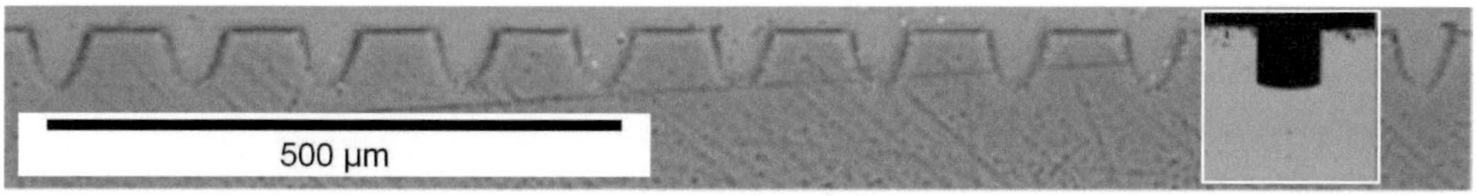

Figure 3.2.: Cross-sections of channels fabricated with a femtosecond laser into stainless steel, source: H. Besser / M. Mangang (KIT IAM); comparison with a channel of comparable dimensions etched into silicon (small picture)

Micromilling

Two methods which are better capable of coping with material restrictions are laser ablation and micromilling. The latter works in analogy to its macroscopic counterpart, yet uses very high precision machines and cutting tools with nominal diameters well below 100 µm. The smallest cutter diameter available in the context of this work is 10 µm with a nominal tool tip radius of approximately 5 µm. The micromilling tool in theory produces a microchannel with cross-sectional shape identical to the tool's geometry. Consequently, one expects the channels' cross-section to be half-circular, elliptical, or "u-shaped" as shown in figure 3.3. In peripheral cutting, also triangular shapes are expected. Measurements reveal a geometrical deviation from the nominal shape, partly to a considerable extent. This is attributed to technology limits in producing perfect tool shapes of such small size, tool wear, and limitations in the cutting process itself.

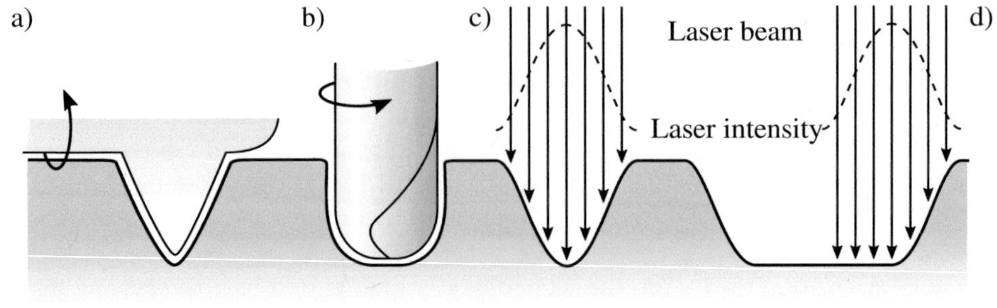

Figure 3.3.: Cross-sectional geometry of a microchannel as a consequence of fabrication (simplified illustration): a), b) micromilling, c) laser ablation d) laser ablation with multiple lines of operation

Laser ablation

Laser ablation is even less dependent on material restriction than micromilling – it machines virtually any given material. Material removal lasers typically focus light energy to a small point on the surface of the material. There, the material heats, melts and evaporates, sublimes, or at best changes directly over to plasma state leaving a dip in the surface [200]. The desired ablation pattern is created by a scanner which moves the laser spot over the material. A microchannel, for example, can be produced in a simple line movement. In order to minimize the heat affected zone and possible bulgings at the channel sides, cf. figure 3.4, it is preferable to have the available light energy concentrated into extremely short pulses so that the actual laser power is highest. Test pieces used in next section's experiments were machined with pico- and femtosecond lasers. Besides pulse length, the quality of the lasered structure depends on an array of tunable parameters such as focus position, pulse overlap, or scanner speed.

The expected cross-sectional geometry in laser ablation is a Gaussian function as depicted in figure 3.3. As the intensity profile of a laser beam follows a Gaussian distribution it is the basic assumption that such a profile is mapped into the channel geometry. This basic understanding does not hold any more in more detailed analysis [89, 106] which underlines the importance of measurement. Commonly, one channel is produced by multiple material removal (movement) lines to create the desired channel depth. In lateral direction, narrow channels with $20 \lessapprox b \lessapprox 40$ µm, are machined in a single line operation, while wider channels allow for multiple parallel removal lines. As a consequence, the cross-sectional shape is expected to become more trapezoidal, as seen in figure 3.3d).

Figure 3.4.: Microchannels fabricated by femtosecond laser ablation into a) stainless steel, b) polyoxymethylene (POM), source: H. Besser / M. Mangang (KIT IAM)

Micromolding and laser interference metallurgy

For completeness of the picture, two more microfabrication technologies shall be mentioned briefly, which especially have the potential for industrial application in higher quantities: One is micromolding [126, 213], the miniaturization of a common molding process, and the other is laser interference metallurgy [62, 93]. Here, microchannels are not created subsequently by a moving laser spot but simultaneously as the result of a laser interference pattern. According to experiments conducted, the depth of microchannels fabricated with the current state of research is not yet sufficient to produce a strong capillary force.

Measurement methods

To determine the cross-sectional geometry, microchannels were cut, metallographically prepared and analyzed by SEM as depicted for example in figure 3.2. Such destructive testing proved superior to microscopic measurements: Confocal laser scanning microscopy and white light interferometry both struggled with absorbed and scattered light especially in deep channels with considerable surface roughness, yet delivered more reliable results with channels of $b > 50$ µm and $\Lambda \ll 1$ (see figure 3.5). As tactile and optical surface roughness measurement possibilities are limited, roughness was judged by SEM images.

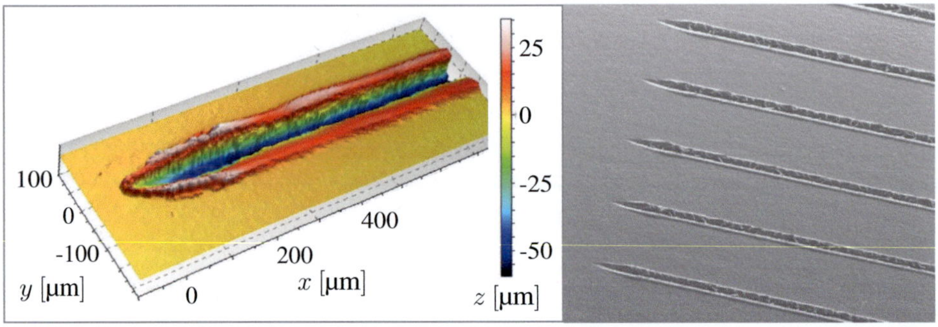

Figure 3.5.: Microchannels of approximately triangular cross-section fabricated by peripheral milling into polyamide 6-6; SEM image (right); geometry measurement (left) via confocal microscopy (M. Egner, Bosch)

Results

Laser ablation and micromilling proved generally capable of generating microchannels in tribological materials. Channel widths as small as $b = 20$ µm have been successfully fabricated, yet surface roughness and geometrical shape improved significantly towards $b = 30$ µm. In the investigated range up to $b = 80$ µm, the aspect ratio was limited to $\Lambda \lessapprox 1$. With respect to surface roughness, the best results were achieved in laser ablation of steel, which resulted in smooth channel walls while roughness at the channel bottom could not be eliminated with holes from discrete laser pulses still visible as in figure 3.4 a). Even though surface roughness may be significant, most channels allowed to distinguish micro- from macrogeometry.

Measurement of cross-sectional shape of the channels confirmed – at least in approximation – the basic assumption that the laser intensity profile or the cutter shape are mapped in the channels' geometry. The expected channel profiles shown in figure 3.3 were fabricated with limited possibility for desired deviation based on machining parameters. Indeed, a suitable set of parameters only allowed for sufficient geometric quality in channel cross-section with a counterexample shown in figure 4.2. Consistency of cross-sectional geometry both along a channel and from one channel to another is not superb, see figures 3.2 and 3.4. The former shows that the laser process, on which most conducted experiments rely (see next section), tends to deviate from a Gaussian cross-section towards a slightly wider, nearly elliptical shape.

Bulgings at both sides of the channel may appear both in milling as in laser machining. Although not critical for lubricant transport, tribological application must be checked for possible conflicts. At least the absence of sharp edges – both at the top and at the bottom of the channels – reduces concern about a possible notch effect.

Consequences for experiments and modeling

Disregarding roughness, the assumption of constant cross-section along the channel still seems approximately valid. The cross-sectional shape is clearly a function of the process and must be obeyed in modeling. Figure 3.3 shows possible geometries. Especially the rectangular cross-section, as the probably most widely covered type in literature, is not amongst them. Even though for later experiments channels with better quality are used, some geometrical uncertainty remains. Measurement allows to reduce such uncertainty, yet not completely due to limitations in nondestructive measurement combined with the fact that results from destructive testing do not perfectly apply to other channels. Instead of relying on a specific shape later modeling must investigate shape variations. Also the experimental design must be adapted.

3.3. Experiments

3.3.1. Isothermal flow experiments

Microscale flow rate measurements

In section 3.1 the flow rate was identified as a critical parameter for the supply of a tribocontact. On the microscale, direct flow measurement is not trivial. A variety of measurement principles have been reported, based on mechanical part movement like in a flow turbine, fluid pressure drop or weighing with measurable flow rates as low as 20 nl/s [8]. Actual flow rates are expected and computed to be lower, for smaller geometries $q < 1$ nl/s. This is not reached by anemometric flowmeters either which measure the flow induced cooling of a heated element. If flow is driven by capillary forces any geometry alteration may change flow characteristic significantly due to changing menisci. Direct force measurement is even less trivial than measurement of flow rate.

As a consequence, it is common practice to determine the flow rate and the forces indirectly and use fluid velocity $u(t)$ as the quantity to be measured. In capillarity driven flow, this is typically done by videotaping meniscus position as a function of time, cf. [4, 71, 103, 212]. Then by means of suitable mathematical models (see next chapter) the quantities of interest are computed. Figure 3.6 shows a front of advancing oil in parallel capillaries with images taken at different points in time. If additionally the flow profile in the microchannel is of strong interest, particle image velocimetry (PIV) and related methods can be applied [150, 214, 218, 219]. Here, tracer particles are added to the fluid, irradiated with a special light source, commonly a laser, particle movement is tracked by a video camera and special software then computes a two dimensional, in some cases even three dimensional vector field of the flow. As microchannel geometry is comparably simple (straight, constant cross-section, unidirectional flow), literature agrees well on microscale boundary conditions (see section 2.4), and adding particles may always influence flow itself especially in small channels, the basic approach is chosen here. Computed flow profiles are later compared to experimental data in literature. For future research, the application of particle tracing techniques might be interesting especially when geometry and flow fields are more complex e.g., as at the end of the microchannel in the vicinity of the tribocontact, cf. chapter five.

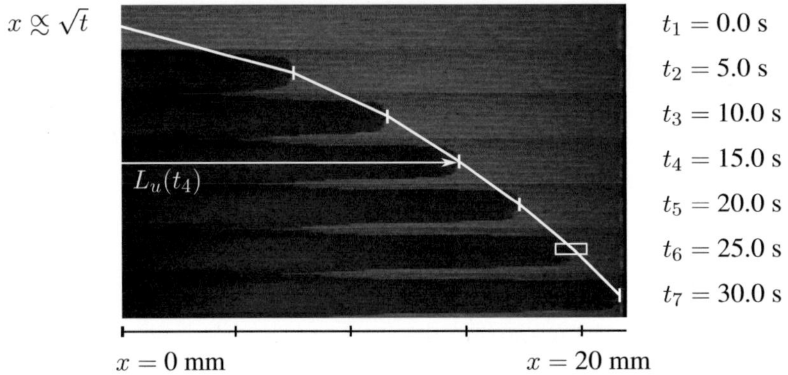

Figure 3.6.: Advancing oil front in multiple parallel channels; images taken at seven time points reveal Washburn-like behavior; the white rectangle indicates the location of figure 3.8 (taken in a different experiment); oil feed out of a brush (instead of using a reservoir) leads to a later start of the capillary flow at the sides of the main oil front

Test pieces and test liquids

For preliminary tests, parallel microchannels were cut into different solid materials, which are considered as characteristic in tribological applications, such as steels 100Cr6, X5CrNi18-10, and X90CrMoV18 as well as engineering plastics polyamide (PA) 6-6, polyphtalamide (PPA), acrylonitrile butadiene styrene (ABS), polyoxymethylene (POM) homo- and copolymer. Test results indicated that differences in materials could primarily be explained by known or measurable material properties, first of all contact angle. In an attempt to reduce experimental variety, only two of the above materials, namely hardened stainless steel X90CrMoV18 and POM homopolymer, were used in subsequent test series, cf. [142]. For the POM, test piece surface was left untreated to keep the smooth surface from the injection molding process [94] whereas steel was polished.

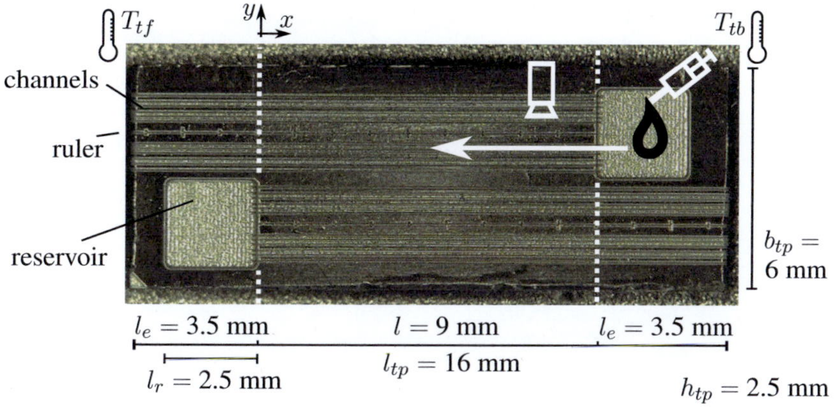

Figure 3.7.: Test piece with a total of 24 straight microchannels

Microchannels were cut into test pieces with dimensions given in figure 3.7. Two channel groups, each consisting of 12 straight channels, were fabricated. Preferred method was laser ablation for its flexibility. Some similar micromilled pieces were used for comparison. Channel length was kept constant at 13.5 mm, while channel width and height were varied. From fabrication all channels had an open surface and cross-sectional geometry as described in the previous section.

Four different greases' base oils were used as experimental lubricating fluids. Table 3.1 lists properties measured in [29] at ambient (NTP) conditions. Density and viscosity were determined on a viscometer (SVM3000, Anton Paar, Austria) while surface tension was measured on a tensiometer (K11, Krüss, Germany) which relies on the Wilhelmy plate method. Equilibrium contact angle measurement was done on the same supplier's drop shape analyzer DSA10. Base oil PAO ISO VG 15 without additives was used in the majority of experiments conducted, with the other oils reserved for plausibility checks. Lubricating fluid influence is considered of higher importance than solid material influence. It would therefore be desirable to refine and extend the conducted experiments' results by testing a larger variety of fluids under different conditions as described in the following.

Test execution

Before conducting experiments, the test pieces were cleaned with isopropanol and petroleum ether both manually and in an ultrasonic bath. Ambient temperature was logged.

Table 3.1.: Measurement of physical properties of four PAOs used in experiments; data given as listed in [29] for NTP conditions, see figure 3.11b) for measurement values at different temperatures

		ISO VG 15	ISO VG 32	ISO VG 68	ISO VG 220
η_n	[Pa s]	$32.4 \cdot 10^{-3}$	$116.6 \cdot 10^{-3}$	$145.1 \cdot 10^{-3}$	$575.9 \cdot 10^{-3}$
ρ_n	$[\text{kg/m}^3]$	816	986	832	844
γ_n	[N/m]	$29.6 \cdot 10^{-3}$	$31.6 \cdot 10^{-3}$	$29.9 \cdot 10^{-3}$	$30.3 \cdot 10^{-3}$
$(\theta_{eq})_n$	[°]	3.9	12.0	10.8	11.6
$\cos\left((\theta_{eq})_n\right)$	[–]	1.00	0.98	0.98	0.98

The microchannels were filled out of a fluid reservoir, cf. [38, 51], which was equally cut into the test pieces, see figure 3.7. With dimensions of 2.5 mm by 2.5 mm and a depth of $\geq$ 200 µm it is both significantly deeper and has a large volume compared to the channels. It was filled by the help of a small volume syringe (Nanofil, World Precision Instruments, USA). Once the fluid level in the reservoir reaches the bottom of the channels, capillary flow starts, at best simultaneously for all channels without the disturbing influence of capillary flow in the reservoir's edges. Reservoir structuring at right angle to channel direction and a slight overlap of channel and reservoir structures were beneficial [217]. The large size of the reservoir helped to reduce variations between experiments, especially as flow velocity within the reservoir is expected to be small compared to channel flow. The reservoir geometry and filling process proved to be more repeatable than other reservoir shapes such as circular and other filling methods e.g., with brush or clock oilers used in preliminary experiments, cf. [29].

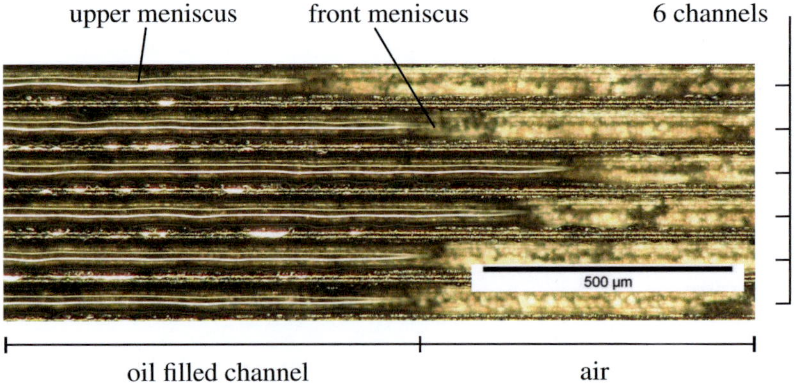

Figure 3.8.: Oil front in six parallel microchannels; microscope focal plane at the upper channel edge with the channel bottom blurred; light reflections give insight into the shape of the menisci

The oil flow is videotaped with a 54 megapixels digital microscope (VHX-600, Keyence, Japan) with additional external light source. For distance measurement, a ruler is lasered into the test piece which allows the readjustment of the field of view during oil flow [142]. The size of the view field is chosen to be big enough to simultaneously observe one of the channel groups and to cover the flow dynamics in the initial flow stage.

Test evaluation

The video taken is analyzed after measurement by noting the travel time for a certain distance. Figure 3.9a) shows time points of the oil advancement distance for three different channel types with mean value and standard deviation. Channel length is defined to be zero at the reservoir with viscous contributions inside the reservoir assumed small due to the above statements. Time is set to zero with oil outflow from the reservoir and timing is taken at every millimeter. For reasons of image quality, videotapes are analyzed manually instead of using image recognition software. The advancement time at a given distance is taken as the mean of the channel group i.e., the mean time of the individual channels μ_t^{ch}.

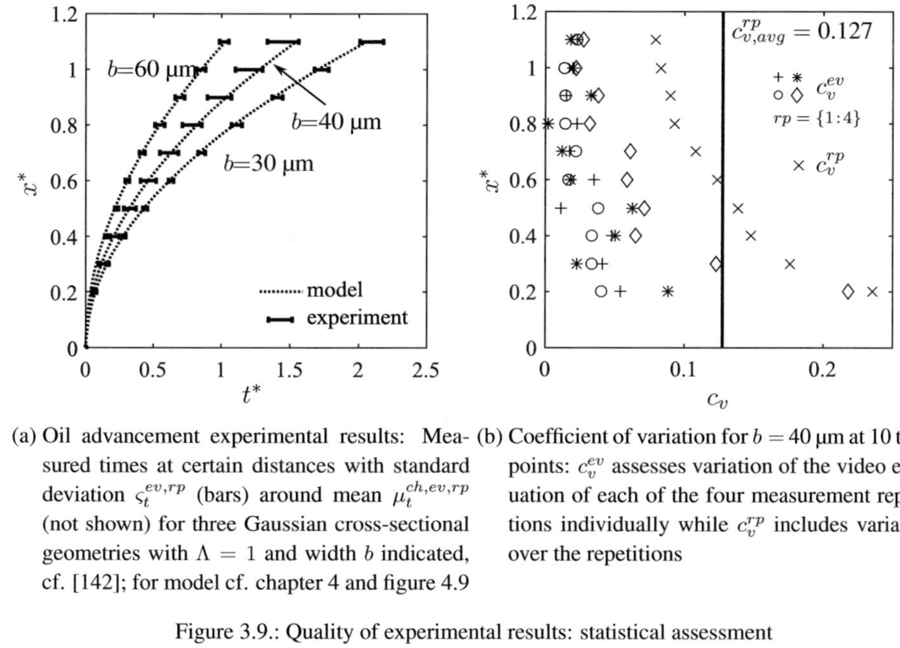

(a) Oil advancement experimental results: Measured times at certain distances with standard deviation $\varsigma_t^{ev,rp}$ (bars) around mean $\mu_t^{ch,ev,rp}$ (not shown) for three Gaussian cross-sectional geometries with $\Lambda = 1$ and width b indicated, cf. [142]; for model cf. chapter 4 and figure 4.9

(b) Coefficient of variation for $b = 40$ µm at 10 time points: c_v^{ev} assesses variation of the video evaluation of each of the four measurement repetitions individually while c_v^{rp} includes variation over the repetitions

Figure 3.9.: Quality of experimental results: statistical assessment

What can be observed is a slightly discontinuous flow: From time to time, individual channel flow stops for a short time which is attributed to channel surface effects: Either due to surface roughness or due to chemically changed surface properties from contamination, the meniscus gets pinned only to advance further milliseconds later. This suggests, yet not proves, that the filling height of the channel is not continuous along the whole channel, a fact further addressed in section 4.3.1, which deals with an upper meniscus.

Each video is evaluated two to three times, in part by two different persons, to reduce bias error. For the evaluations, mean and standard deviation are computed as $\mu_t^{ch,ev}$ and ς_t^{ev}, respectively. As standard deviation is naturally higher for higher travel times, it is normalized for comparison reasons by the respective mean, following the definition of the coefficient of variation c_v:

$$c_v^{ev} = \frac{\varsigma_t^{ev}}{\mu_t^{ch,ev}} \tag{3.1}$$

With the same test piece, multiple experiments (typically four) are conducted as some variation in manual channel filling, test piece cleaning and environmental conditions cannot be ruled out. Again, mean, standard deviation and coefficient of variation are computed over all evaluations of the experimental repetitions i.e., $c_v^{rp} = \varsigma_t^{ev,rp} / \mu_t^{ch,ev,rp}$.

For the channel with highest standard deviation in figure 3.9a) i.e., for the worst case, the coefficient of variation c_v^{rp} is shown in b) along with the individual c_v^{ev} for four experimental repetitions $rp = \{1 : 4\}$. The latter is a measure for the variation in video evaluation while the former includes the variation which comes in when the same experiment is repeated several times. All variation coefficients diminish with growing time as the oil flow slows down which facilitates visual evaluation. Additionally, filling effects are expected to be greater near the reservoir. The fourth experimental repetition deviates considerably from the three other repetitions: While the high $c_{v,rp=4}^{ev}$ at the first time points may be improved by repeated evaluation, the large gap between the individual measurements and the combined c_v^{rp} results mainly from the fact that advancement in the fourth repetition for unknown reasons was considerably slower than in the other three repetitions which leads to the higher standard deviation depicted in figure 3.9a). To compare the overall variation for different channel geometries, an average $c_{v,avg}^{rp}$ over all time points is computed. It is seen in figure 3.9b) as straight line and is listed in table 3.2 also for the other two investigated channel geometries.

Table 3.2.: Average coefficient of variation for experiments conducted with Gaussian-shaped microchannels with aspect ratio $\Lambda = 1$ and width of $b = 30$, 40, and 60 µm

channel width b	30 µm	40 µm	60 µm
$c^{rp}_{v,avg}$	0.037	0.127	0.067

3.3.2. Flow measurement with thermal gradient

Non-isothermal flow experiments

To account for thermal gradients which are the prerequisite for the appearance of e.g., the thermal Marangoni effect, the experimental setup and procedure outlined above are extended as follows: The test piece front face and back face are heated and cooled respectively so that a desirably steep temperature gradient develops inside the test piece. It is therefore contacted by aluminum nibs, each attached to an aluminum block, which can be heated or cooled by a total of four 72 W Peltier elements as seen in figure 3.10.

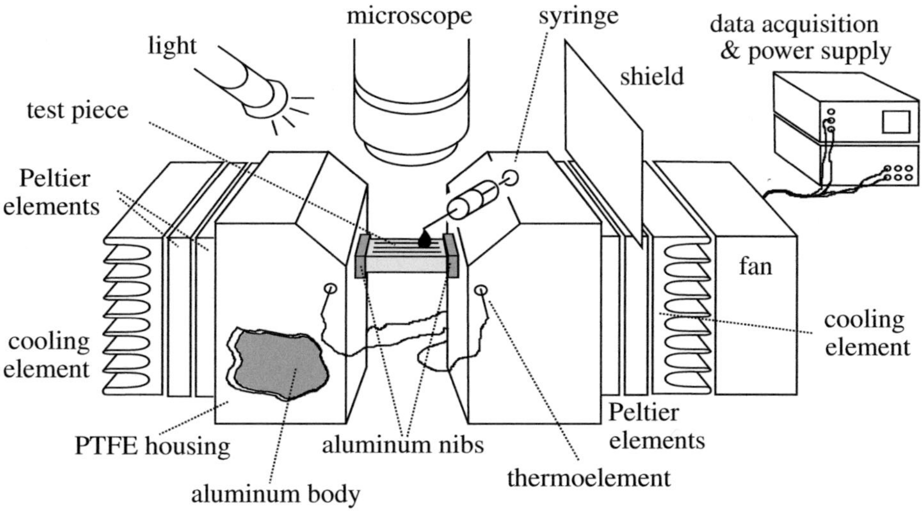

Figure 3.10.: Experimental setup

On the back side of the Peltier elements, cooling elements are mounted and on the side to be cooled an additional fan is installed. To limit convection from the test piece which is surrounded by air on the four remaining faces, a mechanical shield is mounted on the fan. Radiation is reduced by covering the aluminum blocks and nibs with a

PTFE housing. Data of thermocouples inserted into bore holes inside the nibs are used to control the Peltier elements via data acquisition equipment and programmable power supply. The temperature of the test piece itself is checked with thermographic methods. Additionally, calibration test pieces are prepared with three holes to allow for thermocouple measurement in the middle and 2.5 mm away from test piece front and back side, respectively. In this way, the temperature both at the nibs and inside the test piece can be maintained within a close range with standard deviation $\varsigma_T \leq 1.0$ K (POM) and $\varsigma_T \leq 2.8$ K (steel) for the five measurement points per test piece.

Results differ insignificantly from a standard finite element software model, including heat conduction and heat transfer to the surrounding air while radiation is neglected because of the moderate temperature range. For reasons to be discussed in more detail in section 4.4, the model is further simplified to a one dimensional heat transfer model for $T_t(x)$. This again is a good approximation to the three dimensional solution due to the strong enforced gradient in x-direction while lateral and vertical gradients are minor. The simple cuboid geometry allows for an analytical solution similar to the one described in [142]:

$$T_t(x) = T_a + \frac{1}{-2\sinh(B_T l_t)} \cdot \tag{3.2}$$
$$\left(\left[\left(T_{tf} - T_a\right) e^{-B_T l_t} - \left(T_{tb} - T_a\right)\right] e^{B_T x} + \left[\left(T_{tb} - T_a\right) - \left(T_{tf} - T_a\right) e^{B_T l_t}\right] e^{-B_T x}\right)$$

with

$$B_T = \sqrt{\frac{2h^{sa}}{\lambda_s}\left(\frac{1}{b_{tp}} + \frac{1}{h_{tp}}\right)} \tag{3.3}$$

Here, h^{sa} is the solid-air heat transfer coefficient, λ_s is the thermal conductivity of the test piece material, T_{am} is the ambient temperature and the remaining variables are given in figure 3.7. A model fit via h^{sa} to measurement data is shown in figure 3.11a).

While the experimental setup can be used both for metal and plastic test pieces, the focus is on the latter as the more critical material: Its much lower thermal conductivity produces a larger thermal gradient in the vicinity of a tribocontact compared to a metal which contributes more to the removal of heat generated within the tribocontact. The temperature at the experiment's hot side is limited by the glass transition temperature of the test piece and is set to a maximum of 145 °C, whereas a minimal temperature of 5 °C is chosen at the cold side to prevent ice crystals from forming.

The experimental procedure is as outlined above. The two channel systems with one reservoir near the front and one reservoir near the back side now allow two directly subsequent experiments with presumably very similar experimental conditions (except for the channels themselves and the way the oil is dispensed), yet with positive temperature gradient in one case and negative gradient in the other case. This is both relevant once Marangoni and viscosity influences are analyzed in more detail in section 4.4. There, experimental results are presented and directly compared with the model.

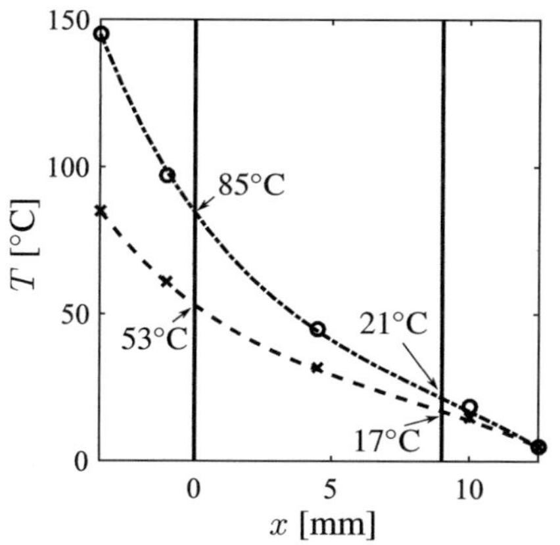

(a) Temperature distribution $T_t(x)$ in POM test piece: measurement values (markers) and theoretical model (lines), cf. eq. (3.2); two variants which differ in hot side temperature (145°C vs. 85°C); the vertical lines indicate the flow domain for use in experiments, cf. figure 3.7

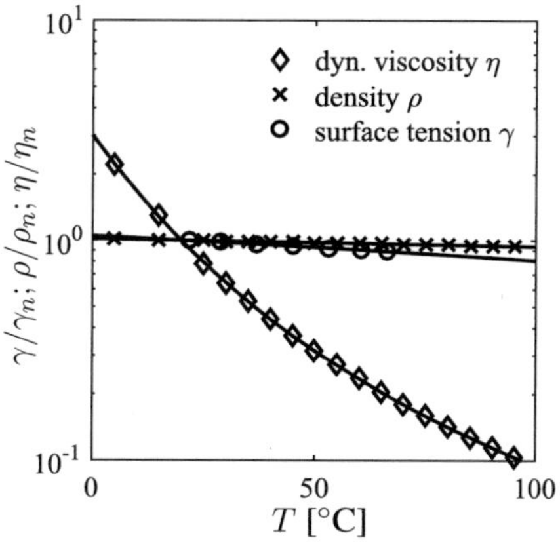

(b) Lubricant properties as a function of temperature; absolute values of PAO ISO VG 15 used in experiments normalized to respective property value at 20°C; measurement data (markers) and computational fit (solid line) via linear models for density and surface tension and via the Ubbelohde-Walter equation (2.7) for dynamic viscosity

Figure 3.11.: Test piece temperature distribution and temperature dependence of lubricating liquid properties, see also [142]

Measurement of temperature dependence of physical properties

As outlined in chapter two, a significant influence of temperature on relevant lubricating liquid properties is expected. Reliable literature data on viscosity's, density's and surface tension's temperature dependence are hardly available for PAOs of interest. It is therefore measured on the equipment named above, cf. [142]. For PAO ISO VG 15, figure 3.11b) shows the measured values normalized to the base value at 20 °C.

4. Microchannel flow model design and results

In the previous chapter, the microchannel array has been identified as a favorable surface geometry to transport lubricating liquids towards a tribocontact. Experiments have been conducted to analyze such a flow. This chapter aims to model the lubricant flow under the typical conditions found in the vicinity of a tribocontact. This requires a slight deviation from the typical Poisson equation commonly used for microchannel flow. Section 4.1 derives a governing set of equations by a nondimensionalization procedure. Section 4.2 discusses geometrical restrictions and argues why the more costly computation of the velocity profile is preferred to simpler concepts for the judgement of viscous friction. Capillary effects, concerning both menisci at the front and upper interfaces of the liquid lubricant column are in the center of section 4.3, which is followed by a discussion of thermal influences: Here the temperature dependent viscosity and the Marangoni effect are major influencing factors. Section 4.5 deals with inertial effects which arise when the microchannels are structured into parts moving relative to fixed space. Based on all previous effects, sections 4.6 and 4.7 illustrate consequences for pressure and especially time dependent behavior of the lubricant flow. The latter builds the basis to compare numerical simulation with results from the previous chapter's experiments. Additionally, the theoretical deviation from classical Washburn behavior is analyzed.

4.1. Derivation of a governing set of equations

Basic assumptions

For two reasons the following analysis does not start with the common microchannel ansatz, eq. (2.28) and (2.29): First, it does not automatically cover the situation where the channel does not constitute an inertial frame of reference in itself but is moving relative to such a frame. Yet, as is described in section 4.5 in detail, the inclusion of body forces could heal this. Secondly, the derivation of Stokes (and Navier-Stokes) equations in the common forms, eq. (2.26) and (2.20), requires viscosity to be constant. With considerable temperature differences present in the vicinity of a tribocontact and bearing in mind the large influence of temperature on viscosity this assumption cannot be upheld here. Therefore, Cauchy's momentum equation in the form of (2.45) is taken as the starting point for the following derivation:

$$\rho \frac{\mathrm{D}\boldsymbol{u}}{\mathrm{D}t} = -\nabla p + \nabla \cdot (\eta (\nabla \boldsymbol{u} + \nabla \boldsymbol{u}^{\mathsf{T}})) + \boldsymbol{f} \tag{4.1}$$

It is valid in a potentially moving frame of reference. For simplicity of reading, the frame indices from (2.45) have been dropped so that $\boldsymbol{u}$ is the fluid velocity relative to the (possibly non-inertial) channel-fixed reference frame (O) as depicted in figure 2.6 which also is the frame time derivatives are computed in (if not otherwise noted). The respective inertial forces ensuring the correctness of the above equation in any frame of reference are included in $\boldsymbol{f}$, cf. section 2.3.4.

In eq. (4.1) simplifications based on the incompressibility assumption, eq. (2.11), and the lubricating liquids' Newtonian behavior which allows to simplify the stress deviator tensor to $\boldsymbol{\tau} = \eta \left(\nabla \boldsymbol{u} + \nabla \boldsymbol{u}^T \right)$, eq. (2.5), have already been considered. For the lubricants intended to flow in the microchannel, especially lubricating oils or base oils from lubricating greases, such assumptions are in general considered safe, cf. section 2.2. Prerequisite are common thermal operating conditions so that the temperature influence on density remains small (see section 4.4 for details). As mentioned, viscosity is not necessarily a constant here due to its extremely strong temperature sensitivity as seen in figure 3.11b). Consequently, eq. (4.1) cannot be further reduced to Navier-Stokes momentum equation (2.20).

Characteristic length scales

The following simplifications are based on procedures used in lubrication theory, cf. section 2.3.3. Yet, they are not adopted in the form used for thin film problems: Whereas there one characteristic length scale is significantly smaller than the other two, for a microchannel two characteristic length scales are significantly smaller than the other one:

$$y_0, z_0 \ll x_0 \tag{4.2}$$

Not only height but also width of the channel is small compared to length, cf. figure 4.1. This assumption should be valid for a vast majority of microchannels discussed in literature. For the channels described in the experimental chapter, length is in the lower millimeter range and therefore greater than width or height by two to three orders of magnitude.

The definition of a small parameter $\delta := (h_0/l_0) \ll 1$ can be adopted from eq. (2.35) where $l_0 := x_0$ and $h_0 := z_0$. In contrast to classical thin film theory, the characteristic lateral channel dimension is not defined as in eq. (2.31) but as

$$y_{0(mc)} := z_0 \tag{4.3}$$

Equal characteristic length scales in y- and z-directions – in analogy to classical theory – do not constitute a loss of generality.

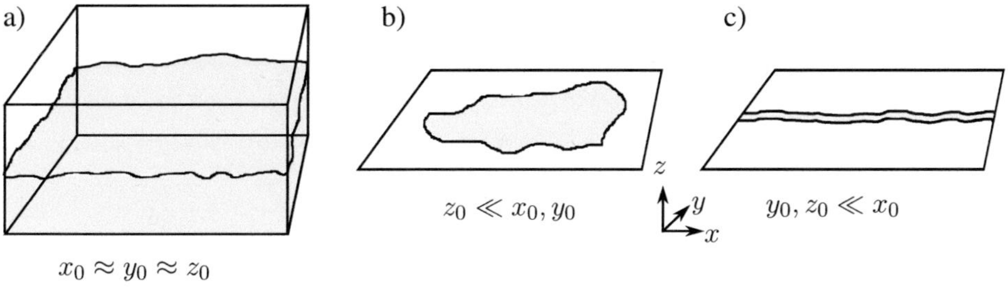

Figure 4.1.: Relation of characteristic length scales in a) bulk fluid, b) thin film, and c) microchannel

Nondimensionalization procedure in analogy to lubrication theory

Based on the above, a nondimensionalization procedure suited for a microchannel can be set up similar to the one used in the derivation of Reynolds equation e.g., by [92]. The individual nondimensionalization of variables is not different from classical lubrication theory as outlined in section 2.3.3 and listed in the appendix, with the exception of lateral coordinate $y^* = y/h_0$ and velocity $v^* = v/\delta u_0$. Viscosity variations require $\eta^* = \eta/\eta_0$. As the key to the analysis is in different length scales, the three equations for the three coordinates have to be treated separately, resulting in lengthy equations. With equal length scales in lateral and vertical directions, the $z-$direction is not printed in the following. Then eq. (4.1) can be written in x-direction:

$$\begin{aligned}
&\rho\left(\frac{u_0^2}{l_0}\frac{\partial u^*}{\partial t^*}+\frac{u_0^2}{l_0}u^*\frac{\partial u^*}{\partial x^*}+\frac{\delta u_0^2}{\delta l_0}v^*\frac{\partial u^*}{\partial y^*}+\frac{\delta u_0^2}{\delta l_0}w^*\frac{\partial u^*}{\partial z^*}\right)\\
&=-\frac{\eta_0 u_0 l_0}{h_0^2 l_0}\frac{\partial p^*}{\partial x^*}+\eta_0\eta^*\left(\frac{u_0}{l_0^2}\frac{\partial^2 u^*}{\partial x^{*2}}+\frac{u_0}{\delta^2 l_0^2}\frac{\partial^2 u^*}{\partial y^{*2}}+\frac{u_0}{\delta^2 l_0^2}\frac{\partial^2 u^*}{\partial z^{*2}}\right)\\
&+\frac{\eta_0}{l_0}\frac{\partial\eta^*}{\partial x^*}\left(\frac{u_0}{l_0}\frac{\partial u^*}{\partial x^*}+\frac{u_0}{l_0}\frac{\partial u^*}{\partial x^*}\right)+\frac{\eta_0}{\delta l_0}\frac{\partial\eta^*}{\partial y^*}\left(\frac{\delta u_0}{l_0}\frac{\partial v^*}{\partial x^*}+\frac{u_0}{\delta l_0}\frac{\partial u^*}{\partial y^*}\right)\\
&+\frac{\eta_0}{\delta l_0}\frac{\partial\eta^*}{\partial z^*}\left(\frac{\delta u_0}{l_0}\frac{\partial w^*}{\partial x^*}+\frac{u_0}{\delta l_0}\frac{\partial u^*}{\partial z^*}\right)+f_0 f_x^*
\end{aligned}$$

in y-direction (z-direction analogously):

$$\begin{aligned}
&\rho\left(\frac{\delta u_0^2}{l_0}\frac{\partial v^*}{\partial t^*}+\frac{\delta u_0^2}{l_0}u^*\frac{\partial v^*}{\partial x^*}+\frac{\delta^2 u_0^2}{\delta l_0}v^*\frac{\partial v^*}{\partial y^*}+\frac{\delta^2 u_0^2}{\delta l_0}w^*\frac{\partial v^*}{\partial z^*}\right)\\
&=-\frac{\eta_0 u_0 l_0}{\delta h_0^2 l_0}\frac{\partial p^*}{\partial y^*}+\eta_0\eta^*\left(\frac{\delta u_0}{l_0^2}\frac{\partial^2 v^*}{\partial x^{*2}}+\frac{\delta u_0}{\delta^2 l_0^2}\frac{\partial^2 v^*}{\partial y^{*2}}+\frac{\delta u_0}{\delta^2 l_0^2}\frac{\partial^2 v^*}{\partial z^{*2}}\right)\\
&+\frac{\eta_0}{l_0}\frac{\partial\eta^*}{\partial x^*}\left(\frac{u_0}{\delta l_0}\frac{\partial u^*}{\partial y^*}+\frac{\delta u_0}{l_0}\frac{\partial v^*}{\partial x^*}\right)+\frac{\eta_0}{\delta l_0}\frac{\partial\eta^*}{\partial y^*}\left(\frac{\delta u_0}{\delta l_0}\frac{\partial v^*}{\partial y^*}+\frac{\delta u_0}{\delta l_0}\frac{\partial v^*}{\partial y^*}\right)\\
&+\frac{\eta_0}{\delta l_0}\frac{\partial\eta^*}{\partial z^*}\left(\frac{\delta u_0}{\delta l_0}\frac{\partial w^*}{\partial y^*}+\frac{\delta u_0}{\delta l_0}\frac{\partial v^*}{\partial z^*}\right)+f_0 f_y^*
\end{aligned}$$

With the definition of Reynolds number $\mathrm{Re} = (\rho u_0 l_0/\eta_0)$ and reduced Reynolds number $\mathrm{Re}_r = \delta^2\mathrm{Re}$, eq. (2.36), the disregard of small terms of order δ^2 and higher leads to the following three equations:

$$\mathrm{Re}_r\left(\frac{\partial u^*}{\partial t^*}+u^*\frac{\partial u^*}{\partial x^*}+v^*\frac{\partial u^*}{\partial y^*}+w^*\frac{\partial u^*}{\partial z^*}\right) \tag{4.4}$$

$$=-\frac{\partial p^*}{\partial x^*}+\eta^*\left(\frac{\partial^2 u^*}{\partial y^{*2}}+\frac{\partial^2 u^*}{\partial z^{*2}}\right)+\frac{\partial\eta^*}{\partial y^*}\frac{\partial u^*}{\partial y^*}+\frac{\partial\eta^*}{\partial z^*}\frac{\partial u^*}{\partial z^*}+\frac{\delta^2 l_0^2}{\eta_0 u_0}f_0 f_x^* \tag{4.5}$$

$$0=-\frac{\partial p^*}{\partial y^*}+\frac{\delta^3 l_0^2}{\eta_0 u_0}f_0 f_y^* \tag{4.6}$$

$$0=-\frac{\partial p^*}{\partial z^*}+\frac{\delta^3 l_0^2}{\eta_0 u_0}f_0 f_z^* \tag{4.7}$$

Further simplification possibilities

Already at this step, velocities perpendicular to downstream direction (i.e., v and w) drop out of the equations: The whole problem is reduced to unidirectional flow. Even though this might be perceived as physically clear – already from inspection of a long and slender channel – it involves direct further simplification in the continuity equation:

$$\nabla \cdot \boldsymbol{u} = \frac{\partial u}{\partial x} = 0 \tag{4.8}$$

Even though not formally required since, the above entails that the microchannel in question must have constant cross-section along x in order to not violate unidirectionality. Disregarding surface roughness (treated in section 4.2.4), chapter three concluded that this is indeed (approximately) the case.

The remaining terms including δ^2 have not been neglected in equations (4.4) - (4.7), because there are factors which might counterbalance those terms when sufficiently large i.e., $\gtrapprox (1/\delta^2)$ or $\gtrapprox (1/\delta^3)$, respectively. The first such possible counterbalancing factor is the Reynolds number itself as part of Re_r. Reynolds numbers in microchannel flow – independent of driving mechanism being a capillary meniscus or some other externally applied pressure – show wide variation, yet upper bounds become apparent: In their overview article Koo and Kleinstreuer [119] list maximum Reynolds numbers researchers have reported of up to 4000 while Steinke and Kandlikar [179] report $\mathrm{Re} \leq 1 \cdot 10^4$ in over 150 publications, provided hydraulic diameter is less than 300 µm. Therefore, in most cases the reduced Reynolds number Re_r is negligible. This is especially true for pure capillary flow, where Reynolds numbers alone tend to be low, often less than 1. As a direct consequence time dependent and convective terms can be disregarded. This has further implications for flow dynamics: As in the derivation of the classical Lucas-Washburn equation (2.71) a quasi-stationary ansatz in the flow time function is justified; it is discussed in more detail in section 4.7.

The other terms which include δ^2 depend on externally applied forces which are theoretically not bounded. For brevity, two more dimensionless factors are introduced:

$$\xi := \frac{\delta^2 l_0^2}{\eta_0 u_0} = \frac{h_0^2}{\eta_0 u_0} \tag{4.9}$$

$$\varkappa := \xi f_0 \tag{4.10}$$

Another possibility for further simplification arises with respect to viscosity. Later analysis will discuss circumstances which allow the reduction of its dependence on three dimensional space $\eta(\boldsymbol{x})$ to just the longitudinal channel coordinate $\eta(x)$. If this is the case together with the remarks above, eq. (4.4) - (4.7) simplify to:

$$\frac{\partial p^*}{\partial x^*} = \eta^* \left(\frac{\partial^2 u^*}{\partial y^{*2}} + \frac{\partial^2 u^*}{\partial z^{*2}} \right) + \varkappa f_x^* \tag{4.11}$$

$$\frac{\partial p^*}{\partial y^*} = \delta \varkappa f_y^* \tag{4.12}$$

$$\frac{\partial p^*}{\partial z^*} = \delta \varkappa f_z^* \tag{4.13}$$

In addition to the negligibility of crossflows v and w, lateral and vertical pressure gradients are significant only in the presence of strong inertial and/or body forces in the respective direction. If such forces are negligible because $\delta\varkappa \ll 1$, the absence of pressure gradients in direction of small geometrical scales (y and z) is in fine analogy to thin film lubrication theory where $\partial p / \partial z = 0$, eq. (2.37). Whereas in thin film theory the remaining two nonzero equations can be integrated and combined to give Reynolds equation in the end, here the absence of forces directly reduces the original three equations required to solve the problem to only one central equation (4.11).

Yet, inertial forces may be relevant as will be discussed in section 4.5, especially in downstream direction: Compared to the other two coordinate directions they are relatively more significant in eq. (4.11) by a factor of $1/\delta$. If still small, eq. (4.11) reduces (after changing back to dimensional quantities) to the basic microchannel flow equation (2.28), provided that viscosity can be treated constant as well.

A more general case which still allows for comparably simple solution needs additional restrictions on the force densities: While the x-component of an external force $f_x(x)$ is allowed to vary in channel direction but not perpendicular to it, the lateral and vertical force components f_y, f_z are assumed to be independent of x or even constant in space. Motivation for such restrictions arises from a combination of typical application conditions and mathematical favor by allowing a strict separation of variables based on the assumption of unidirectional flow. Treated in more detail below these restrictions are the basis for a large reduction in computation time. The loss in generality, which is considered bearable, is discussed in section 4.5. With this, switching back to dimensional quantities yields:

$$\frac{1}{\eta(x)}\left[\frac{\partial p(x,y,z)}{\partial x} - f_x(x)\right] = \frac{\partial^2 u(y,z)}{\partial y^2} + \frac{\partial^2 u(y,z)}{\partial z^2} \tag{4.14}$$

$$\frac{\partial p(x,y,z)}{\partial y} - f_y = 0 \tag{4.15}$$

$$\frac{\partial p(x,y,z)}{\partial z} - f_z = 0 \tag{4.16}$$

From eq. (4.15) and (4.16) pressure is a linear function in vertical and lateral directions along the whole microchannel. Taking pressure to be known at a certain point $p(\boldsymbol{x} = \boldsymbol{X})$, commonly a point at channel inlet where pressure is either known to be the hydrostatic pressure of a large fluid reservoir or simply atmospheric pressure, then with the help of a characteristic, maybe average pressure in channel cross-section $p_m(x) = p(x, y{=}Y, z{=}Z)$ the pressure field is defined by

$$p(x,y,z) = p_m(x) + f_y(y - Y) + f_z(z - Z) \tag{4.17}$$

Characteristic Poisson type flow equation

Even under the presence of external forces (which are subject to the above discussed restrictions) the flow problem reduces in essence to one single Poisson type equation, which governs the unidirectional microchannel lubricant flow:

$$\frac{1}{\eta(x)}\left[\frac{\partial p_m(x)}{\partial x} - f_x(x)\right] = \frac{\partial^2 u}{\partial y^2} + \frac{\partial^2 u}{\partial z^2} \tag{4.18}$$

This equation resembles the classical microchannel flow equation (2.28) except for the fact that external body forces and viscosity are allowed to be nonconstant along the channel. Constant inertial and/or body force components perpendicular to the channel flow direction are equally covered. The common solution method via a separation ansatz, cf. [27], is still possible: In the stationary case i.e., for a given fluid column length, the strict separation of variables requires both sides of eq. (4.18) to be equal to a constant C. In the quasi-stationary case, fluid column length L_u depends on time t so that $C = C(L_u(t))$.

This is is the enabler for a split-up in two equations:

$$\frac{1}{\eta(x)}\left[\frac{\partial p_m(x)}{\partial x} - f_x(x)\right] = C(L_u(t)) \tag{4.19}$$

$$\frac{\partial^2 u(t)}{\partial y^2} + \frac{\partial^2 u(t)}{\partial z^2} = C(L_u(t)) \tag{4.20}$$

The first equation is used to compute C given knowledge about viscosity, external forces and pressure boundary conditions. It essentially determines the magnitude of the downstream flow (surface effects neglected). In the base case of a fully filled microchannel section of length L_u in an inertial frame with known difference between inlet and outlet pressures $\Delta p = p_i - p_o$, constant viscosity and in the absence of body forces,

$$C(L_u(t)) = -\frac{\Delta p}{\eta L_u(t)} \tag{4.21}$$

What can already be seen in the inversely proportional behavior $C \propto 1/L_u$ is that as the fluid column gets longer C declines in absolute value and the flow slows down. This is in line with Washburn's equation and will be discussed in more detail in section 4.7. Up to then, time dependence is neglected.

Having determined C with the help of eq. (4.19), the second of the two equations, eq. (4.20), is used to compute the cross-sectional downstream velocity field $u(y, z)$. From the above assumption of constant channel cross-section it is equally constant along the channel – thus allowing fast computation.

4.2. Geometrical influence on resistance to lubricant flow

For the ultimate goal, the lubricant supply of a tribocontact, the velocity profile inside the microchannel is not directly relevant: What counts is the amount of lubricant transported per unit time irrespective of the fact how fast different fluid layers travel. This section illustrates how the flow rate is significantly influenced by microchannel boundaries which justifies exact velocity profile computation instead of simpler methods to judge flow resistance. Numerical and analytical solution methods are compared: While the former allow for more flexibility in channel geometry (see for example figure 4.2), the latter prove advantageous both with respect to computational cost as well as regarding later analytical analysis in timely behavior in section 4.7.

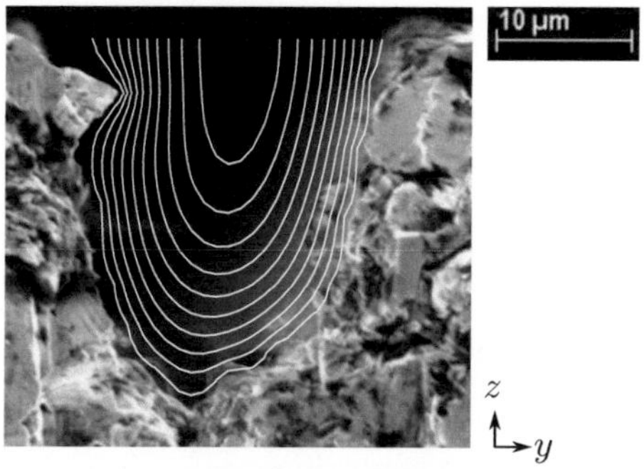

Figure 4.2.: Overlay of cross-sectional velocity profile $u(y, z)$ (contour lines shown) and SEM image of channel lasered into stainless steel with undesired irregular boundary

4.2.1. Cross-sectional channel boundaries

Channel wall boundary condition

Due to fabrication reasons the microchannels investigated here are exclusively open channels cut into a solid substrate, cf. section 3.2. For a non-moving solid channel wall boundary, the no slip boundary condition in the form of eq. (2.47) applies. This is still true for moving structured parts as the boundary movement is to be seen relative to the channel frame of reference, cf. section 2.3.4. Even though relative boundary movement may occur in a tribocontact where two solid bodies can have different velocities $\boldsymbol{u}_A$ and $\boldsymbol{u}_B$ as seen in Reynolds equation (2.38), this is not the case for a microchannel in the surroundings of tribocontact where there is only one solid body. Prerequisite is that solid boundaries can be considered inelastic, both in tangential and normal directions. Again, while elastic deformation is decisive within the tribocontact itself (it is integral part of EHD theory), outside the contact deformations are negligibly small for the investigated engineering materials, because forces are small: Pressure for example is expected to be in the range of atmospheric pressure or below. As slip length does not have to be considered on the microscale (as opposed to the nanoscale, cf. section 2.4.2), the no slip boundary condition in the above mentioned form is applied for all solid walls in the following.

Macroscopic channel shape is discussed in the next section while the surface roughness section 4.2.4 deals with microscopical channel wall geometry.

Open interface boundary conditions

At an open channel's upper interface the fluid-fluid boundary between the lubricant inside the channel and the air above it requires both normal and tangential stress components to be considered. The former may result in a capillary meniscus at the upper interface and are therefore discussed in the following section 4.3. Unlike these, tangential stresses are vital in velocity profile computation. The tangential stress balance eq. (2.59) applies and is further simplified based on the minute viscosity of air, $\eta_A(\text{air}) \approx 18 \cdot 10^{-6}$ Pa s at NTP conditions relative to the lubrication liquid's in the channel (experimental PAOs feature dynamic viscosities of $\eta = \eta_B(\text{oil}) = 32 \cdot 10^{-3}$ .. $576 \cdot 10^{-3}$ Pa s as listed in table 3.1). This allows the use of eq. (2.60).

If the lubricating liquid's stress vector at the interface $\boldsymbol{\tau}_B \cdot \boldsymbol{n}$ can be assumed to point in channel direction only (as will be motivated in section 4.3.1), then for unidirectional flow u parallel to the interface eq. (2.60) simplifies further to yield:

$$\frac{\mathrm{d}u}{\mathrm{d}\boldsymbol{n}} = \frac{1}{\eta} (\nabla_S \gamma) \cdot \hat{\boldsymbol{e}}_x \tag{4.22}$$

The only cause for a surface tension gradient in this work is the Marangoni effect. Marangoni stress τ_M (with the directional index dropped from eq. (2.65) for the simplifying assumption above) consequently determines the velocity gradient at the open channel interface. If one rules out the surfactant-based Marangoni effect, the presence of a temperature gradient determines the appropriate boundary condition:

$$\frac{\mathrm{d}u}{\mathrm{d}\boldsymbol{n}} = \begin{cases} \tau_M/\eta & \text{(non-isothermal)} \\ 0 & \text{(isothermal)} \end{cases} \tag{4.23}$$

The second, isothermal case constitutes the classical "no stress" boundary condition.

4.2.2. Velocity profile computation

Given such boundary conditions, the governing equation for velocity profile computation, eq. (4.20), in principle still allows for an analytical solution, yet only for a few special basic cross-sectional shapes like triangular [139, 167], rectangular (see below) and trivially for circular cross-sections. The latter results in the classical parabolic velocity profile for laminar flow in pipes or tubes as seen in figure 2.2b). Flow in other channel geometries may still be calculated analytically on the basis of conformal mapping [154] or perturbation theory [27]. Alternatively, numerical methods can be applied to cover non-trivial cross-sections with curved open interface. As the present work is not primarily focused on solution methods for a well-known type of differential equation, a brief discussion of one analytical solution approach and one numerical ansatz follow.

Numerical solution for arbitrary channel shapes

Out of different possible numerical approaches a finite differences ansatz is presented in the following mainly for its simplicity: The basic method to solve the discrete Poisson equation is readily available in literature, e.g. in [85]. Instead of curvilinear coordinates [111, 182] a rectangular grid is chosen with grid points $i = 1..n$ in y-direction and $j = 1..m$ in z-direction, respectively. This leads to

$$\frac{\partial^2 u}{\partial y^2} = \lim_{\Delta y \to 0} \frac{1}{\Delta y} \left(\frac{u_{i+1,j} - u_{i,j}}{\Delta y} - \frac{u_{i,j} - u_{i-1,j}}{\Delta y} \right) = \lim_{\Delta y \to 0} \frac{u_{i-1,j} - 2u_{i,j} + u_{i+1,j}}{\Delta y^2} \tag{4.24}$$

with $\partial^2 u / \partial z^2$ determined analogously. The numerical approximation results by not considering the limit but instead a finite grid spacing Δy. As the channel aspect ratio is roughly $\Lambda \approx 1$ due to fabrication reasons, the grid spacing can additionally be equidistant so that $\Delta y = \Delta z$. This allows to write eq. (4.20) in discrete form as the equation system:

$$\frac{1}{\Delta y^2} \left(u_{i-1,j} + u_{i,j-1} - 4u_{i,j} + u_{i+1,j} + u_{i,j+1} \right) = C \tag{4.25}$$

Grid points coming to lie outside the flow domain are excluded from the computation. No slip (Dirichlet) boundary conditions at a channel wall are covered by considering the respective value (e.g., $u_{i,0} = 0$ at the channel bottom) in the equations which determine the fluid velocity adjacent to that wall. Stress (Neumann) boundary conditions require ghost points located beyond the boundary: The stress on the boundary, for example at $j = m + 1$, is numerically approximated in a central difference scheme by

$$\frac{\partial u_{i,m+1}}{\partial z} = \frac{\tau_M}{\eta} \approx \frac{u_{i,m+2} - u_{i,m}}{2\Delta y} \tag{4.26}$$

which serves to compute the ghost point values $u_{i,m+2}$. With them, the equations to solve for the velocities on the boundary at $j = m + 1$ read:

$$\frac{1}{\Delta y^2} \left(u_{i-1,m+1} + 2u_{i,m} - 4u_{i,m+1} + u_{i+1,m+1} \right) = C - \frac{2\tau_M}{\eta \Delta y} \tag{4.27}$$

In case of a Marangoni effect being present at a non-flat upper surface, the use of eq. (4.26) to compute the surface stress with respect to the z-direction (instead of the normal direction as eq. (4.23) would require) results in a slight underestimation of some velocities $u_{i,j}$ especially near the upper channel edges. The relative error is bounded above by the relative length increase of the curved compared to the flat upper meniscus ($s_o^{\perp}/b - 1$), yet is expected to be significantly smaller, below one per cent, so that it is neglected in the following. In general, algorithms like Gauss-Seidel or SOR can be applied to speed up the solution process.

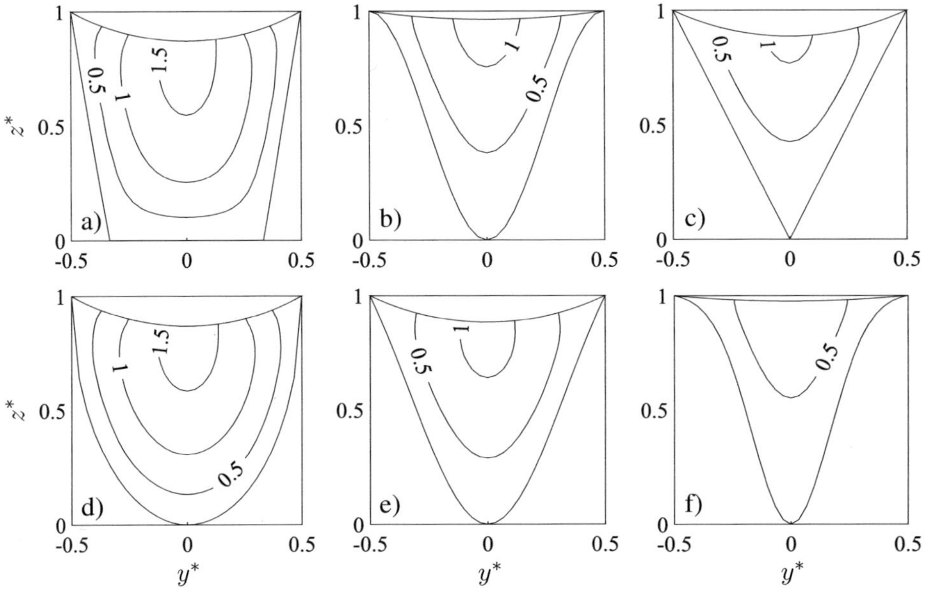

Figure 4.3.: Velocity profiles $u^*(y^*, z^*)$ in different channel cross-sections: a) trapezoidal, b) sine, c) groove (v-shape), d) elliptical (u-shape), e) wide Gaussian (3ς), f) narrow Gaussian (6ς); for the rectangular cross-section see figure 4.5; comparison based on equal C, i.e., equal front meniscus pressure drop, channel length, and constant viscosity; for differences in upper meniscus see section 4.3

Results are shown in figure 4.3 for cross-sectional shapes which are expected as possible outcome of fabrication (cf. section 3.2). All cross-sectional geometries share equal C (i.e., equal capillary pressure, channel length and constant viscosity). Upper meniscus curvature has been considered, yet still without interfacial stress.

Definition of the input geometry

The most suitable input geometry to calculation is defined by measurement on real channels. Even though any measured shape can be computed, cf. figure 4.2, production variation along a channel, between channels, from material removal methods and due to the nature of the machined material leads to the requirement of defining a characteristic range for likely cross-sections. In section 3.2 the elliptical and Gaussian shapes depicted in figure 4.3d) and e) have been identified as bounds for such a range – at least with respect to the channels used in the experiments which are to be compared with the model. All following computations are based on these two geometries with the Gaussian shape defined as the standard geometry if not otherwise noted. Its theoretical shape is constructed as flipped normal distribution probability density function with zero mean and standard deviation $\varsigma = 3$ with tails cut to fit aspect ratio of $\Lambda = 1$.

If channels are fabricated in a different way (e.g., by milling, multiple line laser removal operations, etc.) a respective characteristic shape has to be defined.

Semi-analytical solution for rectangular cross-sections

Channels with a close to perfectly rectangular cross-section can be produced by etching, cf. figure 3.2 (inlet), while milling and laser ablation still allow for such a shape with acceptable geometrical exactness at least in the case of larger microchannels with width and depth in the higher micrometer range. To solve the governing Poisson equation on a rectangular domain, different types of Fourier series approaches are pursued e.g., by [26, 103, 151, 218]. The following ansatz is set up in consideration of the stress boundary condition at a (flat) open channel interface as outlined above, eq. (4.23):

$$u(y,z) := \sum_{n=1}^{\infty} f_n^F(z) \cdot \cos(g_n^F y) \tag{4.28}$$

where $g_n^F = (2n-1)\pi/b$. Key to the solution is the decoupling of the two geometry variables by creating a product of z-dependent coefficients $f_n^F(z)$ and a series of y-dependent trigonometric functions. The no slip boundary conditions on the left and right channel walls are forced into the equation by defining a simplified rectangular function with

$$S^F(y) := \begin{cases} C & \left]-{}^{b}/_{2}\,,\,{}^{b}/_{2}\right[\\ 0 & \text{elsewhere} \end{cases} \tag{4.29}$$

Here,

$$S^F(y) = C \cdot \sum_{n=1}^{\infty} k_n^F \cdot \cos(g_n^F y) \tag{4.30}$$

where Fourier coefficients $k_n^F = -4(-1)^n/(g_n^F b)$. Replacing the right hand side in eq. (4.20) with S^F leads together with eq. (4.28) to a second order ordinary differential equation for $f_n^F(z)$. The nature of the top channel boundary (solid wall or open boundary with or without applied stress) determines the solution. In case of an open boundary it reads:

$$f_n^F(z) = \frac{k_n^F}{g_n^{F2}} \left[C\left(1 - e^{-g_n^F z}\right) + \frac{e^{-g_n^F h}\left(e^{g_n^F z} - e^{-g_n^F z}\right)}{1 + e^{-2g_n^F h}} \cdot \left(\frac{\tau_M g_n^F}{\eta} - Ce^{-g_n^F h}\right)\right] \tag{4.31}$$

Boundary conditions transform to $f_n^F = 0\big|_{z=0}$ (no slip) and $(\mathrm{d}f_n^F/\mathrm{d}z)\big|_{z=h} = k_n^F \tau_M/\eta$ (stress).

With respect to the infinite sum in both equations (4.28) and (4.30) some care in computing may be beneficial: At both jump discontinuities in (4.29) the Gibbs phenomenon produces overshoots when only a finite number of terms up to $n = N$ is considered, even if N is large. Sigma approximation is a possible workaround: A cardinal sine function $\mathrm{sinc}(n/N) = \sin(n\pi/N)/(n\pi/N)$ is introduced as additional factor, so that the above mentioned equations read in approximation:

$$u(y,z) \approx \sum_{n=1}^{N} \mathrm{sinc}\left(\frac{n}{N}\right) \cdot f_n^F(z) \cdot \cos(g_n^F y) \tag{4.32}$$

$$S^F(y) \approx C \cdot \sum_{n=1}^{N} \mathrm{sinc}\left(\frac{n}{N}\right) \cdot k_n^F \cdot \cos(g_n^F y) \tag{4.33}$$

Figure 4.4 compares such a semi-analytical approach with the numerical approach described above by the 2D velocity profile in the vertical channel center plane at $y^* = 0$. N is to be chosen in a tradeoff of computational speed and accuracy with $N = 20$ used as standard value.

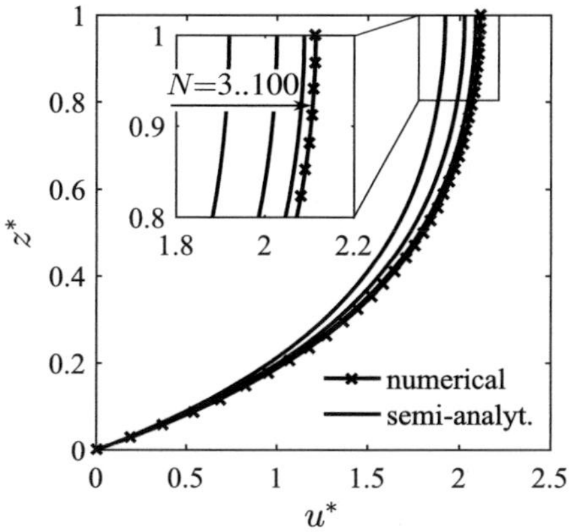

Figure 4.4.: Comparison of semi-analytical and numerical computation; the former based on sigma approximated Fourier series, eq. (4.32), the latter being a finite differences approach, eq. (4.25) and (4.27)

4.2.3. Viscous flow resistance and flow rate

The velocity profiles in figure 4.3 suggest that the channel shape has considerable influence on flow resistance. The absence of a channel cover (owed to production technology) expectedly proves advantageous with respect to viscous friction: A closed channel's characteristic full-parabolic flow profile, cf. figure 2.2b), is significantly slower than a comparable open channel half-parabolic velocity profile as depicted in figure 4.4. Yet the latter allows an inward-curved upper meniscus which constrains the flow domain, cf. figure 4.5a) and b).

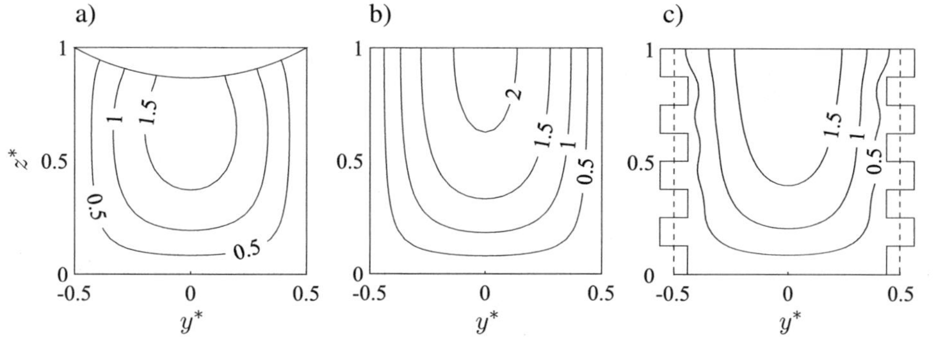

Figure 4.5.: Influence of changes in cross-sectional area on flow profile $u^*(y^*, z^*)$: a) by upper meniscus, c) by idealized surface roughness features in comparison to b) unrestricted flow area of open channel; absolute values comparable with figure 4.3

By integration of the velocity profile over the cross-sectional area, the flow rate q is determined based on eq. (2.50) at any given time t as:

$$q\left(C\left(L_u\left(t\right)\right)\right) = \iint_{(A)} u\left(y, z, C\left(L_u\left(t\right)\right)\right) \mathrm{d}y \, \mathrm{d}z \tag{4.34}$$

From continuity, the flow rate is spatially constant along the whole fluid column as is average downstream flow speed through the cross section $\bar{u} = q/A$. Now, transportation capabilities of the different cross-sections can either be compared directly by q or by related concepts as a normalized Q or hydraulic resistance J_H. Comparison requires constant C i.e, parameters channel length, front meniscus pressure drop and viscosity have to be fixed at a certain value. As comparison standard geometry, the three sigma Gaussian shape, cf. figure 4.3e) is chosen so that

$$Q := q/q^{(3\varsigma)} \tag{4.35}$$

Table 4.1 judges the flow resistance of different channel cross-sections based on various comparison concepts. Cases (2) – (9) correspond to the respective velocity profiles in figures 4.3 and 4.5. Significant differences become obvious, for example the open square channel without upper meniscus, case (2), is able to transport almost eight times as much lubrication liquid compared to the six sigma Gaussian channel, case (9), even though front meniscus pressure drop and channel length are set to equal value. Elliptical and 3ς Gaussian channels still differ by a factor of almost two, underlining the considerable degree of geometrical uncertainty.

Table 4.1.: Various measures of viscous friction in channels with the following cross-sections: (1) square (closed), (2) square (open, flat meniscus), (3) square (open, curved meniscus; same for all following channels), (4) trapezoidal, (5) sine, (6) groove, (7) elliptical, (8) Gaussian (3ς), (9) Gaussian (6ς)

shape	(1)	(2)	(3)	(4)	(5)	(6)	(7)	(8)	(9)

Q	2.62	4.31	2.98	2.09	1.03	0.65	1.96	1.00	0.55
$J_H/J_H^{(3\varsigma)}$	0.38	0.23	0.34	0.48	0.97	1.53	0.51	1.00	1.82
d_H^*	1.00	1.33	1.21	1.12	0.83	0.76	1.14	0.90	0.68
$d_H/d_H^{(3\varsigma)}$	1.11	1.48	1.35	1.24	0.95	0.84	1.28	1.00	0.76

The widely used but criticized concept of a hydraulic diameter d_H (see section 2.4.2 for more details) deviates from the exact computation of the flow rate via velocity profiles. The latter is consequently preferred in the following in spite of the computation simplicity of d_H (which may still serve as a rough indicator) and the concepts which build on it e.g., friction factors. The hydraulic resistance is inversely proportional to the flow rate and therefore equally an appropriate measure.

Deviations from a straight microchannel course

The geometrically simple straight microchannel analyzed until now is not universally applicable. Curves, bends, turns, or corners may be required to reach the tribocontact. In analogy to macroscopic hydrodynamics they increase viscous friction. On the microscale the increase is generally characterized as "moderate". If the channel radius of curvature R_c is large compared to the channel's width

$$\frac{b_c}{R_c} \ll 1 \tag{4.36}$$

the channel can be treated as straight [202].

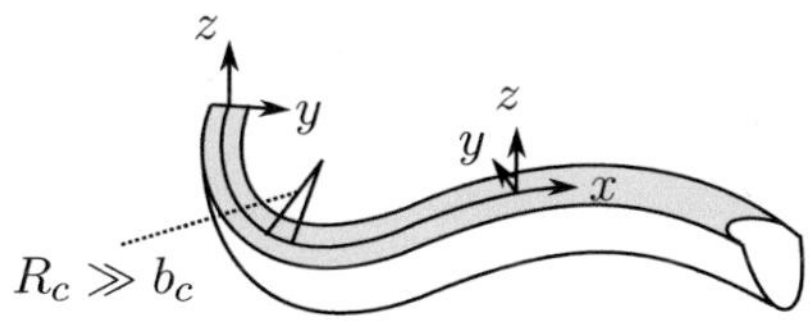

Figure 4.6.: Slight channel curvature based on eq. (4.36) allows the use of curvilinear coordinates

Capillary effects, dealt with in more detail in the next section, may potentially add to flow resistance in non-straight channels. Yet while meniscus induced deceleration before a turn and subsequent acceleration after the turn can be observed, no additional flow resistance contribution for turning angles up to 75° is reported [38]. As capillary effects are less likely to occur in smooth curves than in sharp turns, eq. (4.36) is the condition which decides if the curvature can be neglected and the microchannel behaves like its straight counterpart. In this case the Cartesian coordinate system used so far can be generalized to a curvilinear one where x follows the channel course and y and z keep their orthogonal directions as seen in figure 4.6.

4.2.4. Surface roughness

The numerical approach presented above already allows to consider idealized surface roughness features in the channel cross-section as part of the overall geometry. Figure 4.5c) shows the influence of post-like roughness similar to models of [18, 46, 40, 121].

For small relative roughness ϵ_{rel}, eq. (2.57), such procedure leads to fine mesh requirements and consequently high computational cost: It seems advisable to separate roughness as microgeometrical feature from the channel's macrogeometry. What is not covered yet is the surface roughness in longitudinal direction as it conflicts with the constant cross-section assumption ($\partial u/\partial x = 0$) along the channel. It is included in the constricted diameter approach, cf. section 2.4.2, which is reported valid for $\epsilon_{rel} \leq 14\%$ [110]. Channel roughness, measured from SEM images, is generally below such thresholds for the materials and fabrication methods investigated. Highest roughness comes from holes due to pulsed laser ablation, as seen in figure 3.4a). In contrast to an averaged diameter as seen in figure 4.5c), the constricted diameter excludes such features from flow computation which seems reasonable. Consequently, constricted channel width b_c and depth h_c are taken as the relevant input parameters for computation instead of their nominal counterparts b and h, respectively.

In addition to viscous friction capillary effects must be kept in mind e.g., when the front meniscus travels past roughness spikes and posts. Based on Lattice-Boltzmann modeling Kusumaatmaja et al. [121] report heavy fluctuations in downstream fluid velocity when the meniscus passes an obstacle. This short time behavior converges to smooth Washburn-type advancement for the characteristic timescale t_0 which is based on channel length. Both observations are confirmed in the experiments conducted. A capillarity-induced slowdown of flow in rough channels is covered experimentally in constant G_{ex} introduced in the next section. Theoretically, certain surface patterns are able to stop capillary flow completely by meniscus pinning, cf. figure 3.1e) and [70, 222]. Yet to be effective, such patterns need sharp edges without smaller bypass structures which are highly improbable to occur in a statistically rough surface. Nevertheless, redundancy by multiple channels may be required from a design point of view.

4.3. Capillary effects

Whereas the influence of channel geometry on viscous flow resistance can be determined with good accuracy based on the methods and arguments presented in the previous section, its influence on capillary effects is more intricate. The following discussion especially falls short in providing an exact determination of the capillary force driving the capillary flow exclusively by computational methods. To a limited degree experiments are necessary as described in section 4.3.3. To be able to critically assess the experimental results and to get a more precise understanding – qualitatively and quantitatively – of the capillary effects present, different approaches pursued in literature are discussed in section 4.3.1, focusing on the specific case of liquid lubricant transport towards tribocontact. The meniscus at the upper channel interface may not be neglected in this context: Even though it is less relevant than the front meniscus in driving a capillary flow, such upper meniscus is important in its characteristic to reduce the flow area significantly. Based on this, the potential of different channel cross-sections to build up capillary pressure is computed and compared in section 4.3.2 accompanied by a critical discussion. Conclusions for open channel capillary flow under isothermal conditions are drawn in section 4.3.4.

4.3.1. Modeling and relevance of capillary menisci in given geometries

Open microchannels, especially those with cross-sectional geometry defined in chapter three, are expected to adopt a liquid-air interface shape similar to what is shown in figure 4.7: A smooth shape without sharp border between upper and front menisci. With fluid movement and additional surface stresses due to a Marangoni effect present, the surface cannot be expected to have constant curvature which is the prerequisite for a solution via eq. (2.73) - (2.75). Especially flow dynamics may distort the interface shape thereby altering the pressure drop in normal direction, cf. eq. (2.62). The pressure drop then becomes a function of location on the interface instead of being constant as in Young-Laplace equation (2.63).

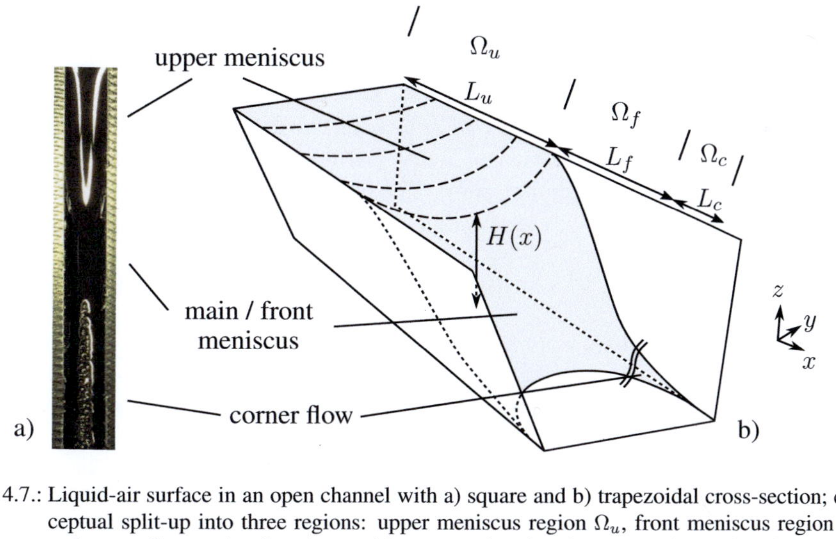

Figure 4.7.: Liquid-air surface in an open channel with a) square and b) trapezoidal cross-section; conceptual split-up into three regions: upper meniscus region Ω_u, front meniscus region Ω_f, and corner flow region Ω_c; picture a) has approximative character only as taken in the static case against gravity in a channel with $b = h = 1000$ µm

A numerical model capable of handling the above requirements well is considered a crucial outcome of future research. One basis may be the work of Brakke [23] which focuses on the minimization of different types of energies and which has been extended by Klatte et al. [116] to include viscous friction via friction factors. A generally promising approach, the criticism associated with friction factor theory, cf. section 2.4.2, applies. Navier-Stokes based interface tracking algorithms, phase field or level set methods may be an alternative. Here, covering the multiple scales from channel length to local interface curvature all the while considering boundary conditions correctly and ensure mass and volume conservation is certainly not straightforward. Molecular dynamics models which are typically applied to nanochannels are being scaled up towards the microscale which is yet computationally costly. In summary, the correct computation of interfacial effects in a given geometry without neglecting bulk flow viscous friction is seen as a challenge, especially considering the different scales a microchannel is characterized by, cf. eq. (4.2).

It is the scaling argument which supports the alternative approach of conceptual split-up of the fluid column into three different regions or flow regimes as depicted in figure 4.7b) in analogy to [146, 187]. As the channel is long and slender, there must be a first region Ω_u which is characterized by upper meniscus curvature yet where slope $\partial H/\partial x$ and its change $\partial^2 H/\partial x^2$, eq. (2.81), are small in channel direction. This is expectedly not true any more at the front meniscus. Based on such reasoning, Ω_u is defined to start at the microchannel inlet at $x = 0$ and continues as far as the channel is filled to the top edge – with an upper meniscus allowed to form. The beginning of the second region Ω_f is marked by the detachment of the contact line from the upper edge. Thereafter, saturation drops rather sharply. This transitional region is identical with the front meniscus. Generally assumed to be small in circular capillaries, this does not need to be the case in other cross-sections. The third domain, the corner flow region Ω_c, is where the liquid fingers may form. The following sections analyze these regions in more detail, especially with regard to the two capillary driving mechanisms, corner and bulk flow:

Liquid finger region Ω_c

Corner flows are expected in region Ω_C in the lower edges of rectangular and trapezoidal channels, as long as the two lower edges are not sufficiently rounded, cf. [163]. In larger channels, where fabrication can meet such requirement, they are well observable as in figure 4.7a).

In contrast, elliptical and Gaussian-shaped channels do not feature such edges but, in analogy to flow in an open v-shaped channel, figure 2.12, can be considered as an edge themselves. As outlined in section 2.5.3, the corner flow in a perfect v-shaped groove is provoked by the ever increasing upper meniscus curvature with diminishing saturation and therefore diminishing lateral meniscus width. Gaussian and elliptical channels exhibit the same behavior, which is yet superposed by varying wall slope angle $\varphi(H_w)$ at different locations along the channel as seen in figure 4.8a):

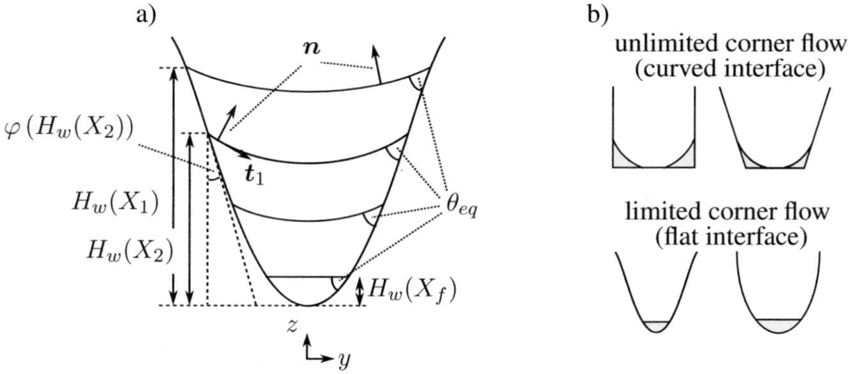

Figure 4.8.: Unlimited and limited corner flow depending on cross-sectional shape

A potential corner flow contribution is maintained as long as $\varphi(H_w(x)) < (\pi - \theta_{eq})$. This is in line with Romero and Yost's [167] condition extended for the fact that $\varphi(H_w(x))$ depends on the height of the contact line $H_w(x)$. It declines along x until at $x = X_f(t)$ the interface becomes flat thus eliminating further corner flow contribution, see figure 4.8. If at all present, liquid fingers are consequently expected to be small in Gaussian and elliptical channels. It is therefore consistent with theory, that such fingers have not been observed in experiments with these channel geometries. From a modeling perspective, the flat interface is the critical condition which rules out corner flow modeling by help of a similarity transformation [167].

In channels with sharp edges where fingers form (e.g., in trapezoidal or rectangular channels) Weislogel [208] has investigated joint corner and bulk flow. Even though his model does not apply to open channels, he basically states that compared to the static case the fingers' importance is less in a flowing liquid. Both volume proportion in fingers as well as the fingers' length are reportedly small, with $L_c \approx 0.03 \cdot L_u$ [103, 208]. In open channels, with only one or two edges, the finger flow contribution is expected to be even smaller. In summary, the contribution of flow in fingers in all investigated cross-sections is expected to play a subordinate role, especially if the respective channel does not feature sharp edges. Finger flow is therefore not further discussed and finger length is defined to be zero, $L_c :=0$.

Front meniscus region Ω_f

With respect to the front meniscus, region Ω_f, Chen [36] experimentally investigates its shape in open trapezoidal channels via high-speed imaging and finds similar meniscus dimensions in x- as well as in z-directions: $L_f \approx h$. Guo et al. [90] report slightly longer front menisci in rectangular channels but still term the transition from the fully-filled channel as "jump-like" so that against the background of characteristic microchannel dimensions $z_0 \ll x_0$ it is reasonable to assume $L_f \ll L_u$ (except maybe for the very short-time after flow start which is out of scope here). Experiments are supported by theoretical analysis: Nilson et al. [146] argue that at the border between regions Ω_u and Ω_f, the local pressure gradient $\partial p/\partial x$ must be both continuous and comparable in absolute value in both regions. With pressure level in Ω_u derived from the upper meniscus balance and the requirement of mean curvature in Ω_u to be in line with front meniscus theory, they arrive at the following formula to estimate the length of the meniscus region:

$$\frac{L_f}{L_u} = \left(\frac{2\gamma}{b_c L_u}\right)^{\frac{1}{3}} \tag{4.37}$$

Even though the front meniscus grows in (absolute) length with the advancing fluid column, its relative length L_f/L_u declines. Valid for an open rectangular channel, this is analogous to the bulk meniscus being compressed to a line in a closed rectangular channel [208]. Only for flow in a v-shaped channel, L_f is reported to be of significant magnitude compared to L_u [167]. From the analysis in the previous section, Gaussian or elliptical shapes are expected to show a behavior in-between trapezoidal and v-shaped channels. In experiments conducted a rather sharp front meniscus is visible e.g., in figure 3.8.

Though not as clear as in the the case of liquid fingers, the front meniscus is equally assumed small in the following, $L_f \ll L_u$. As a consequence capillary flow in the channels under consideration can be modeled in fine analogy to basic microchannel flow models, namely by restricting attention to region Ω_u exclusively while treating the front meniscus with its characteristic pressure drop as a boundary condition.

Upper meniscus region Ω_u

Under the above stated assumptions, the upper meniscus only needs to be considered in a configuration where it is pinned to the top edge as seen in figure 4.7. Then upper meniscus curvature is exclusively determined by the pressure drop across the upper interface, as the contact angle is undefined from Gibbs criterion, cf. figures 2.4c) and 2.13b). Following the basic understanding of capillary flow, pressure inside the liquid column gradually decreases from inlet pressure $p_i = p_{am}$ at $x = 0$ to the underpressure $p_{am} - \Delta p_f$ caused by the front meniscus at $x = L_u$. In other words, the front meniscus pressure drop (together with inertial force contributions as is yet to be seen) is responsible for the pressure distribution in the whole fluid column. As pressure acts in all three spatial directions equally, the upper meniscus pressure drop in each cross-section along the fluid column can then be readily computed as difference across the upper interface:

$$\Delta p_u(x) = p_{am} - p(x, y{=}0, z{=}H) \tag{4.38}$$

This is based on several assumptions: First, Δp_u is a function of x only as the pressure drop is considered constant across the interface line in each cross-section. This in turn assumes Young-Laplace equation (2.63) to be valid. In the special case here, where quasi-stationary unidirectional flow is in essence parallel to the interface, eq. (2.62) suggests that such an assumption is much more reliable than in front meniscus and corner flow regions. As a consequence the simplifying cross-sectional definition $p(y, z{=}H_y(y)) := p(y{=}0, z{=}H)\ \forall\, y$, cf. figure 2.13a), must be valid. Only in the presence of external forces f_y,f_z this conflicts with the cross-sectional pressure distribution, eq. (4.17). Requiring more detailed analysis, at least for small forces and small upper meniscus curvature Young-Laplace equation should still be approximately valid. Maintaining the assumption that curvature is negligible in downstream direction, eq. (2.81), the cross-sectional upper meniscus is a circular arc with its radius of curvature given by eq. (2.82):

$$R_u(x) = \frac{\gamma}{\Delta p_u(x)} \tag{4.39}$$

As the difference $\Delta p_u(x)$ grows in channel downstream direction, the upper meniscus radius of curvature diminishes. Figure 4.7 shows the flat interface at channel inlet (provided the common inlet condition $p(x{=}0) = p_{am}$, the pronounced upper meniscus $R_u(x{=}L_u)$ and two exemplary transitory meniscus shapes in between. At $x = L_u$, the meniscus detaches from the upper channel edges. It is not governed any more by the Gibbs criterion but by the contact angle, eq. (2.83), where in this case $H_w = h$ and φ is the wall slope angle at the upper channel edge, cf. figure 4.8:

$$R_u(x{=}L_u) = \frac{b}{2\cos(\theta_{eq} + \varphi(H_w{=}h))} \tag{4.40}$$

Together with $R_u(x)$, $S(x)$ and $A_S(x)$ decline within Ω_u in downstream direction in a not necessarily negligible way: Depending on contact angle and channel aspect ratio, $A_S(x{=}L_u)$ may be less than half of $A_S(x{=}0)$, thus causing a modeling contradiction:

As the requirement of flow rate continuity is out of question, the average flow rate across the channel's cross-section $\bar{u}$ must be a function of x:

$$q = \bar{u}(x) \cdot A_S(x) = \text{const} \tag{4.41}$$

This in turn conflicts with the very basic assumption $\partial u/\partial x = 0$, eq. (4.8), much of the previous analysis is based upon. Indeed, diminishing $A_S(x)$ requires a slight flow contribution w in negative z-direction which together with boundary condition $w = 0$ at $z = 0$ requires growing downstream velocity from incompressibility, $\nabla \cdot \boldsymbol{u} = 0$. Still, $\partial u/\partial x \approx 0$ from $x_0 \gg y_0, z_0$, eq. (4.2). The average velocity relative range equals the filled cross-section relative range ($\max \bar{u}/\min \bar{u} = \max A_S/\min A_S$) and is therefore well bounded. As for tribological applications not $\bar{u}(x)$ but constant q is of major importance, the presented model is not abandoned but simplified further: At this point a mean velocity $\tilde{\bar{u}}$ is defined as the velocity which together with a mean average filled cross-section $\tilde{A}_S$ fulfills eq. (4.41). In other words, if the microchannel in consideration had constant saturation with filled cross section $\tilde{A}_S$ it would allow the velocity $\tilde{\bar{u}}$ which would result in the same flow rate q as in the real channel. Table 4.2 visualizes such a model simplification and contrasts it with the expected real flow.

Table 4.2.: Model simplification via averaged filled cross-section $\tilde{A}_S$ instead of $A_S(x)$, visualized for a Gaussian cross-section, with consequences for cross-sectional velocity average $\bar{u}$

	$x = 0$		$x \approx L_u/2$		$x = L_u$	
expected / observed		$\bar{u} < \tilde{\bar{u}}$		$\bar{u} = \tilde{\bar{u}}$		$\bar{u} > \tilde{\bar{u}}$
model		$\bar{u} = \tilde{\bar{u}}$		$\bar{u} = \tilde{\bar{u}}$		$\bar{u} = \tilde{\bar{u}}$

Still, $\tilde{A}_S$ itself has to be determined. From eq. (4.40) upper meniscus curvature is known at $x = L_u$. With the help of eq. (4.39) the respective upper meniscus pressure drop $\Delta p_u(x{=}L_u)$ can be computed. It also allows to determine the upper meniscus radius of curvature $\tilde{R}_u = 2R_u(x = L_u)$ which corresponds to exactly half this pressure difference, $\Delta p_u(x{=}L_u)/2$. This is the basis for the different upper meniscus shapes shown in figure 4.3. Then with the definition

$$\tilde{A}_S := A_S(X)|_{\Delta p_u(X) = \Delta p_u(x=L_u)/2} \tag{4.42}$$

$\tilde{A}_S$ is the cross-section at half distance between the channel inlet and the front meniscus, in case of the typical constant pressure gradient $\partial p/\partial x = \text{const}$.

Criticism

In the presence of inertial forces due to a moving reference frame (which lead to a nonconstant pressure gradient, cf. section 4.6) this is not exactly true any more. The expected modeling errors still have to be characterized.

Additionally, it should be expected that the front meniscus pressure drop Δp_f as determined in the following sections and the maximum upper meniscus pressure drop $\Delta p_u(x = L_u)$ are equal. As this is not necessarily the case, the questions must be raised if the equilibrium contact angle θ_{eq} in eq. (4.40) is indeed the correct contact angle and if the transition between regions Ω_u and Ω_f is in reality not as sharp as modeled.

Furthermore, the presented modeling concentrates on the front meniscus pressure drop as the cause for the creation of an upper meniscus which in turn reduces the filled channel area and therefore the throughput through the channel. What is disregarded is the fact, that the upper meniscus itself might actively influence the pressure distribution in the channel and maybe even alter the front meniscus. A similar reasoning has been brought up by Furrer et al. [75] while discussing gravity-driven open channel flow with an upper meniscus.

In conclusion, simplifying assumptions have been made which still allow for a computationally inexpensive unidirectional flow model. Eq. (4.34) is slightly updated to yield

$$q\left(C\left(L_u\left(t\right)\right)\right) = \iint_{(\tilde{A}_S)} \tilde{u}\left(y, z, C\left(L_u\left(t\right)\right)\right)\, \mathrm{d}y\, \mathrm{d}z \tag{4.43}$$

Complete neglection of the upper meniscus is not seen as an alternative – for the significant reduction in cross-sectional area available for downstream flow.

4.3.2. Theory-based quantitative front meniscus assessment

From basic understanding the front meniscus is responsible for creating the capillary driving force. In section 2.5 two approaches to compute the capillary force have been introduced – the thermodynamic approach, built on the minimization of Gibbs free energy, and the alternative ansatz of interfacial pressure drop. The thermodynamic approach (cf. section 2.5.5) proves as the recent method of choice for open channels, applied e.g., in [15, 16, 36, 151]. It promises to handle the critical influence of channel cross-sectional geometry including an open interface with relative ease.

The geometrical factor G as used by Sammarco and Burns [173] and introduced in section 2.5.2 is chosen as basis to compare computation and experiments. It basically extracts the critical aspect, namely geometry, from the computation of capillary force F_{cap} and pressure drop Δp_f, thereby separating it clearly from other influencing factors such as channel size or physical properties, cf. eq. (2.76).

Comparison requires the capillary force from energy minimization to be converted into an equivalent front meniscus pressure drop:

$$\Delta p_f = \frac{F_{cap}}{\tilde{A}_S} \tag{4.44}$$

Together with eq. (2.76) and (2.85) it is possible to compute G based on the energy minimizing approach as

$$G_{em} = \frac{b_c}{\tilde{A}_S} \left(s_w^{\perp} - \frac{s_o^{\perp}}{\cos\theta} \right) \tag{4.45}$$

Table 4.3 lists values for G_{em} based on a fixed aspect ratio of $\Lambda = 1$ (except for $\Lambda = 0.5$ for the half-circular cross-section). The small contact angle approximation $\cos\theta \approx 1$, eq. (2.16), is applied.

Table 4.3.: Comparison of the magnitude of the front meniscus pressure drop in open channels with different cross-sectional shapes by help of the geometrical constant G_{em} based on eq. (4.45); aspect ratio is $\Lambda = 1$ except for the half circular shape where $\Lambda = 0.5$; for closed channels see table 2.2; for detailed shape description see table 4.1

shape	(1) □	(2) ⊔	(3) ◡	(4) ⊔	(5) V	(6) V	(7) U	(8) U	(9) V
G_{em} (no upper men.)	4.00	2.00	1.45	2.03	2.61	2.47	1.61	2.17	3.24
G_{em} (upper men.)	-	2.15	$\rightarrow 2$	2.21	2.66	2.84	1.98	2.44	3.35

For each channel cross-section, two values for G are given in table 4.3, one for a flat open interface and one including an upper meniscus. In comparison with a flat interface, the upper meniscus elongates the open channel boundary $s_o^{\perp}$ and from eq. (2.85) thereby reduces the overall capillary force. Yet this effect on meniscus pressure drop is outweighed by the reduction of cross-sectional area from A to $\tilde{A}_S$ by the inward curved meniscus.

A mathematically interesting special case is the half circular cross-section with upper meniscus. Especially channels produced in a milling process may feature such a shape. With the small contact angle assumption $\cos\theta \approx 1$, meniscus shape and channel shape are identical so that both capillary force and cross-sectional area vanish. As a consequence, there should be no capillary flow. Yet mathematically $G_{em} \rightarrow 2$ and in experiments flow is observed (even though the used channels lack perfectness in both shape and surface roughness).

Additionally remarkable is the fact that in an open channel as opposed to a closed channel capillary force is not only reduced by shorter wetted perimeter $s_w^\perp$ but additionally by its replacement with a liquid-air interface $s_o^\perp$. This leads to the consequence that an open channel of aspect ratio $\Lambda = 1$ and open surface width b_c can theoretically reach at maximum half the pressure drop of the respective closed channel if an upper meniscus is neglected. With upper meniscus, it is slightly more than half in the case of a square cross-section. The difficulties with the half-circular cross-section described above make it difficult to compare it with the full circular cross-section value from table 2.2.

In conclusion, the above followed thermodynamic approach by means of minimization of Gibbs free energy allows to easily compute the capillary force for the required channel cross-sections and with this it is clearly superior to pressure drop computation from meniscus shape as discussed in the previous section. Yet results are questionable already on a theoretical basis. An alternative is the experiment-based determination of the influence of channel geometry.

4.3.3. Experiment-based quantitative front meniscus assessment

Experiments are conducted in the way described in section 3.3 with channels of various widths, starting from $b_c = 30$ µm and rising to 70 µm in steps of 10 µm. The aspect ratio is kept constant at $\Lambda = 1$. The degrees of freedom laser ablation offers were used to keep the channel cross-sectional shape as similar as possible. Consequently, a single G-value is expected to govern the flow in all channels. From section 3.2, the resulting shape is considered to be neither the 3ς Gaussian geometry depicted in figure 4.3e) nor the elliptical shape in figure 4.3d) but somewhere in between. This is in line with laser ablation theory which expects deviations from a Gaussian shape [89, 106].

In the following, the computation is based on both shapes serving as upper and lower bounds for the actual value of G_{ex}. It is determined by fitting the model to the experimental time function of one specific channel type, here $b_c = 60$ µm. Then this fit is validated with the help of the other channel types' experimental data. Missing details on the computation of the time function are given in the following sections, especially in section 4.7. Figure 4.9 shows the liquid lubricant advancement both as a function of time and as a function of square root of time. Good agreement is reached with $G_{ex}^{3\varsigma} = 3.62$ for a Gaussian shaped channel.

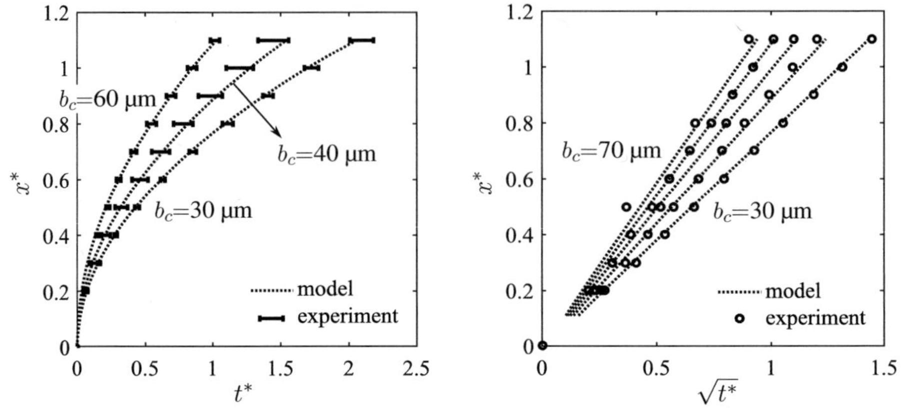

Figure 4.9.: Time function of capillary flow in open channels, $\Lambda = 1$, $b_c = 30$ .. 70 µm; model fit (lines) via $G_{ex}^{3\varsigma} = 3.62$ to experimental data of channel with $b_c = 60$ µm; experimental values of all channels are given with mean value (circles) or standard deviation (horizontal bars); same data as a function of t^* (left) and $\sqrt{t^*}$ (right)

The straight lines in figure 4.9b) indicate – as expected – a perfect Washburn-behavior. If computation is based on an elliptical channel instead, the necessary capillary pressure to meet the measured advancement time function is less, because the elliptical cross-section features less viscous resistance. Table 4.4 contrasts both cases with the theoretical values G_{em} determined in the previous section.

Table 4.4.: Comparison of theoretically and experimentally determined values for G for the upper bound of a Gaussian cross-sectional channel geometry and the lower bound of an elliptical shape, cases (7) and (8) in table 4.3; the last two columns give the respective values in the virtual case when no upper meniscus would be present

	3ς Gaussian (upper bound)	elliptical (lower bound)	3ς Gaussian (upper bound)	elliptical (lower bound)
	with upper meniscus		no upper meniscus	
G_{ex}	3.62	2.64	2.90	2.00
G_{em}	2.44	1.98	2.17	1.61
G_{ex}/G_{em}	1.48	1.33	1.34	1.24

The gap between computed values based on the thermodynamic approach and experimentally derived values is considerable. In the virtual case of no upper meniscus it is only slightly smaller. The same is true for the elliptical channel geometry, but the deviation is still remarkable. Conceptually, the theoretical approach has already been criticized in the previous section. Especially questionable is the fact that observed advancement is not slower but faster than theory predicts – in spite of imperfect real channels. Experimental design as described in section 3.3 allowed to reduce yet not eliminate experimental uncertainty: Besides the stated geometrical uncertainty, unrecognized systematic error could be present in all conducted experiments e.g., due to a significant influence of surface roughness effects or heating due to strong microscope illumination. On the other hand, real-world conditions are possibly covered sufficiently by G_{ex} in a way that repeated experiments cannot disprove computational prediction. For this, the amount and variety of experiments conducted should desirably be increased.

Also with respect to aspect ratio Λ further experiments may help to clarify its influence and potentially analyze its detailed effect on G, similar to the formula of Lenormand et al. [128], eq. (2.77), which is valid for closed rectangular channels.

4.3.4. Conclusions for isothermal flow

Figure 4.9 shows that computation and experiments match well in the isothermal flow case. The special cross-sectional channel geometry which results from fabrication (in this case laser ablation) and the meniscus at the open channel's upper interface have been considered. It is sufficient to determine a single value of geometrical constant G_{ex} by fitting it to any of the experimental curves. It then proves valid for all other investigated channel sizes in the isothermal case (as seen in figure 4.9) and will equally show fine results in the non-isothermal case discussed in the following section. Determining G experimentally is considered safer than theory-based computation as the latter can be questioned with respect to both theoretical consistency and the unsubstantiated way it deviates from measurements.

In general, capillary flow is faster in channels with bigger cross-section. This is true for any given characteristic channel shape and results from the fact that higher viscous friction in small channels overcompensates the increase in front meniscus pressure drop. Obviously, such reasoning is bounded above by the declining relevance of capillarity with size; here capillary length l_{cap}, eq. (2.1), can serve as a useful indicator. While such base case clearly suggests to design channels with large cross-sections for tribocontact supply, the consideration of additional effects as discussed in the following sections puts this reasoning into question – and may well lead to the opposite conclusion.

4.4. Thermal aspects

Thermal aspects in this context are important for the simple reason that a tribocontact generates heat. The introductory section 4.4.1 gives an overview of potential consequences for capillary lubricant transport and discusses how these can generally be implemented into a model. After the influence of temperature dependent physical properties is clarified, a model is developed based again on a nondimensionalization procedure. Special attention is given to the temperature dependence of viscosity in section 4.4.4 and to the Marangoni effect (section 4.4.5).

4.4.1. Tribocontact's influence on lubricant microchannel transport

Frictional power losses in tribocontact can be considerable, if not huge. As already mentioned, about a third of a car's fuel consumption is due to friction [101]. With nearly all of the energy converted into heat, such an amount cannot be neglected. The detailed heat generating mechanisms are out of scope because they happen within the tribocontact and not in its vicinity where the capillary surface structure is located. Consequently, the tribocontact can be modeled simply as a heat source which transfers energy to its surroundings. Potential heat transfer mechanisms include conduction in the solid contact bodies as well as both conduction and advection in moving lubricants while radiation can be neglected for the moderate temperatures encountered.

In the isothermal case analyzed up to now, the capillary transport system has been reduced to the model of a slender microchannel which unidirectionally conducts lubricants from a distant reservoir to the tribocontact. As will be seen, the thermal problem fits such simplification well: The channel begins at a low, maybe ambient temperature inlet, and ends at the hot tribocontact, so that the fluid flow generally faces a positive temperature gradient irrespective of the exact channel course, cf. figure 4.11. This directly implies an opposing force due to the Marangoni effect, cf. section 2.4.5, which tends to draw the lubricant away from the tribocontact.

Of minor concern are phase transitions – condensation and evaporation processes – because lubricants are generally chosen to have boiling points higher than the tribocontact temperature and evaporation processes below the boiling point happen much slower than the characteristic transportation time t_0 of typically a few seconds. Figure 4.10 shows the time function of lubricant flow in a microchannel both in the isothermal case and under positive and negative thermal gradients. The former has already been illustrated in figure 4.9. For the latter, experiments were conducted as described in section 3.3 and computation comes from the model developed in the following. The negative thermal gradient is investigated first for comparison purposes and secondly as one might desire to lead excess liquid away from the tribocontact. Microchannel flow can thereby potentially act as a cooling mechanism transporting cold lubricants towards the tribocontact and warmed lubricants away. It will be briefly discussed why microchannel dimensions yet make this a difficult task.

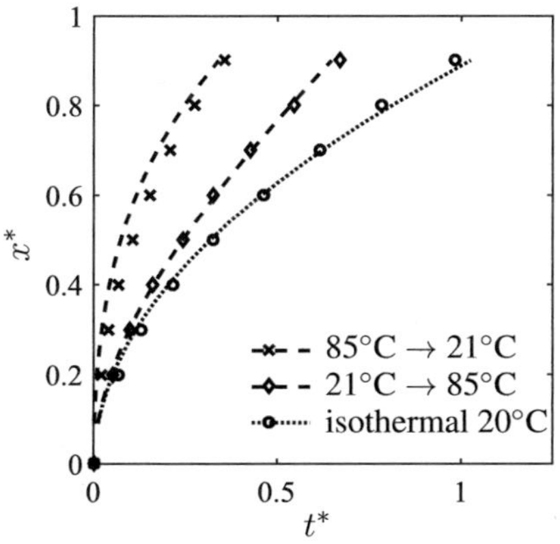

Figure 4.10.: Time function of non-isothermal flow with negative, positive, and zero temperature gradient $\mathrm{d}T/\mathrm{d}x$ in channel with $b_c = 40$ µm; experiments (markers) and model (lines); test piece temperature $T_t(x)$ as displayed in figure 3.11a)

As seen in figure 4.10 both flow with thermal gradient and against it is faster than flow at ambient temperature. This strongly contradicts sole thermally induced Marangoni influence: Compared to the isothermal case one would expect faster flow with Marangoni support but slower flow against a thermal gradient. The result can yet be explained by the temperature dependence of viscosity, cf. section 4.4.4.

The surfactant-based Marangoni effect, cf. section 2.4.5, is considered to be of minor importance in a tribological context: While oil contamination regularly occurs, the presence of surface active agents with relevant concentration gradients is generally improbable. As the mathematical treatment of both types of the Marangoni effect is similar, an extension to oil transport due to surfactants should be easily feasible if required in the special case.

Modeling strategy builds on the assumption that mass and volume of the liquid lubricant column inside the microchannel is generally extremely small compared to the solid body the channel has been cut into. This is based on the fact that the solid body as contact partner does not exclusively act as channel carrier but has to fulfill a variety of mechanical requirements e.g., with respect to mechanical stability, which in turn leads to significantly larger dimensions than the microchannel's height and width. Consequently, fluid temperature's influence on solid body temperature is so small that given sufficient thermal conductivity of the solid, bidirectional thermal interaction can be disregarded. Then modeling can be split up in two stages: At first, the temperature field of the solid body is computed which acts as boundary condition for the determination of the fluid temperature in the second stage. Before this is detailed in section 4.4.3, the influence of temperature on lubricant properties must be further discussed:

4.4.2. Relative relevance of parameters' temperature dependence

Basically all relevant physical properties of lubrication fluids are temperature dependent to a certain degree, cf. section 2.2.2. If the dependence is weak enough, the respective parameter may be treated as constant in a model, thus potentially reducing model complexity and increasing computational speed significantly while result accuracy hardly suffers. With this motivation, the following discussion aims at judging and comparing physical properties' relative temperature dependence. For this purpose, Bruus [27] proposes dimensionless parameters $\mathrm{d}_T\lambda/\lambda$, $\mathrm{d}_T\eta/\eta$, etc.

With regard to surface tension, density, and viscosity the respective temperature dependencies of characteristic PAO lubrication oils have been measured in preparation of the flow experiments, cf. section 3.3 and visualized in figure 3.11b). The respective dimensionless parameters are computed and listed for different temperatures in table 4.5, cf. [97]. They are complemented by respective values for specific heat c_p and thermal conduction λ for similar PAOs based on data given by [9] and computed with the help of eq. (2.12) and (2.13). The latter directly leads to $\mathrm{d}_T\lambda/\lambda = B_\lambda/\lambda_n$.

Table 4.5.: Relative temperature dependence of PAO fluid properties, cf. [97]; viscosity, surface tension, and density from measurement of experimental liquid PAO ISO VG 15 (see section 3.3 and [142]), thermal conductivity and specific heat per unit volume from Bartel [9] for PAO ISO VG 32 (λ) and PAO ISO VG 150 ($c_p \cdot \rho$)

T [°C]	$\frac{\mathrm{d}_T\eta}{\eta}$ [1/K]	$\frac{\mathrm{d}_T\lambda}{\lambda}$ [1/K]	$\frac{\mathrm{d}_T\gamma}{\gamma}$ [1/K]	$\frac{\mathrm{d}_T(c_p\cdot\rho)}{(c_p\cdot\rho)}$ [1/K]	$\frac{\mathrm{d}_T\rho}{\rho}$ [1/K]
20	$-4.75 \cdot 10^{-2}$	$2.39 \cdot 10^{-3}$	$-2.39 \cdot 10^{-3}$	$6.50 \cdot 10^{-4}$	$-7.84 \cdot 10^{-4}$
40	$-3.59 \cdot 10^{-2}$	$2.51 \cdot 10^{-3}$	$-2.51 \cdot 10^{-3}$	$6.42 \cdot 10^{-4}$	$-7.97 \cdot 10^{-4}$
80	$-2.25 \cdot 10^{-2}$	$2.77 \cdot 10^{-3}$	$-2.79 \cdot 10^{-3}$	$6.26 \cdot 10^{-4}$	$-8.23 \cdot 10^{-4}$

On this basis a comparison of the relative strength of temperature dependence of the different fluid properties yields:

$$\left|\frac{\mathrm{d}_T(c_p \cdot \rho)}{(c_p \cdot \rho)}\right| \approx \left|\frac{\mathrm{d}_T\rho}{\rho}\right| < \left|\frac{\mathrm{d}_T\lambda}{\lambda}\right| \approx \left|\frac{\mathrm{d}_T\gamma}{\gamma}\right| \ll \left|\frac{\mathrm{d}_T\eta}{\eta}\right| \tag{4.46}$$

Temperature's influence on viscosity is clearly the strongest. Based on measurements of PAO ISO VG 15 depicted in figure 3.11b) viscosity at 0°C is higher than viscosity at 100°C by a factor of 31.9 – and such temperature range may still not be sufficient for applications. Surface tension is significantly less influenced by temperature. The basis of the Marangoni effect, it yet becomes decisive for very slow flow – a realistic case in the vicinity of a tribocontact. Consequently, viscosity and surface tension are treated as functions of temperature in what follows, while the other parameters' temperature dependence is neglected. Such an assumption is debatable for thermal conductivity, which features a change of approximately 2.5% over a range of 10 K. Specific heat capacity and density are less affected by temperature so that especially the basic incompressibility assumption can be upheld.

Based on the three interfacial tensions as expressed in Young's equation (2.15), contact angle is a function of temperature as well. Bernardin et al. [13] have conducted experiments with water on metal surfaces and found continuously decreasing θ_a between 0°C and 250°C, with higher slope towards higher temperatures. For contact angles of lubricating oils on metal surfaces similar behavior is reported [122]. For oils with low contact angles already at ambient temperature, cf. table 3.1, a further declining contact angle with rising temperature entails minute changes in $\cos\theta$. Thus for such liquids the contact angle can be treated as de facto constant in temperature.

4.4.3. Thermal modeling

Lubricant temperature

Inside the lubricant fluid column, thermal energy transport occurs either by heat conduction or by advection. In downstream direction, in spite of generally low fluid velocities in microfluidics, the thermal Péclet number can be significant:

$$\mathrm{Pé}_{\mathrm{th}} = \frac{l_0 u_0}{\chi} > 1 \tag{4.47}$$

In such a case, advection dominates over energy diffusion in the main liquid flow direction. Yet due to the slender geometry of the microchannel the Nusselt number

$$\mathrm{Nu} = \frac{h^{sl}}{\lambda_l}\frac{4A}{s_w^{\perp}} < 1 \tag{4.48}$$

is typically small indicating the relative importance of conductive compared to convective heat transfer across the solid channel wall: As a consequence, one expects the temperature distribution inside the channel to depend only slightly on the lubricant velocity but decisively on the boundary temperature. Confirmation comes from nondimensionalization in a very similar procedure to the one described in section 4.1 and which is again based on $y_0, z_0 \ll x_0$, eq. (4.2). It starts with the energy equation in the simplified form of eq. (2.22) as from above thermal conductivity is treated as constant. Via

$$\delta^2 \boldsymbol{u}^* \nabla^* T^* = \frac{\chi}{u_0 l_0} \left(\delta^2 \frac{\partial^2 T^*}{\partial x^{*2}} + \frac{\partial^2 T^*}{\partial y^{*2}} + \frac{\partial^2 T^*}{\partial z^{*2}} \right) \tag{4.49}$$

it leads to the Laplace equation

$$\frac{\partial^2 T}{\partial y^2} + \frac{\partial^2 T}{\partial z^2} = 0 \tag{4.50}$$

which governs the temperature distribution in the channel's cross-section. For constant boundary temperature T_w it yields uniform temperature $T = T_w$ inside the fluid. The same is true in good approximation if T_w applies only to the solid channel wall s_w and not to the open channel boundary s_o as long as heat transfer over the liquid-air interface is small relative to the solid-liquid interface – a reasonable assumption.

Solid temperature

As the solid body the microchannel is engraved into necessarily has characteristic dimensions $l_s, b_s, h_s \gg y_0, z_0$ it is indeed fine to assume that the solid body temperature variation between the left, right and lower channel walls at a specific channel cross-section (i.e., over characteristic distance y_0) is minute. Then, the solid wall temperature exclusively depends on x: $T_w(\boldsymbol{x}) = T_w(x)$. In summary, the liquid temperature T equals the solid boundary temperature along the channel, and therefore is a function of downstream location only:

$$T(\boldsymbol{x}) = T(x) = T_w(x) \tag{4.51}$$

From here the solution is straightforward with the computation of the solid's temperature distribution with a single heat source (the tribocontact) and heat sinks at the body's boundaries being a standard engineering problem based on the heat equation.

Model applicability

With respect to lubricant use as a coolant the above stated is an ambivalent result: While lubricant flow principally can be used to transport cold lubricants towards the tribocontact and thermal energy away from it, care has to be taken that the energy exchange happens at the desired location and is not for example immediately transferred to the solid body carrying the microchannel.

Furthermore, the model so far only allows for stationary conditions. The transient thermal behavior after a tribocontact starts or seizes operation or also during operation is not considered. If design is interested in the worst case, a stationary analysis may be fully sufficient. With respect to applications where tribocontacts are in interrupted operation e.g., in low running gears where the stationary case may never be reached, there is a considerable risk of overengineering so that further refining work is needed. To reasonably model such types of tribocontact the additional question is to be addressed if quasi-stationary flow modeling (already in the isothermal case) is sufficient or if short-time dynamic analysis including inertial terms is required.

The restriction of temperature to be a function of longitudinal coordinate x only is no restriction on the general temperature field inside the solid body, even though real cases might prove relatively simple. For example if one of the contact partners is a flat or hardly curved surface with microchannels arranged to supply the tribocontact radially, the problem is similar to the experiments conducted with the cuboid test piece as described in section 3.3. Yet for arbitrary channel courses on arbitrary temperature fields as illustrated in figure 4.11 the model still applies as long as the solid's temperature $T_s(x)$ is known and the requirements for non-straight channels stated in section 4.2.3 are met.

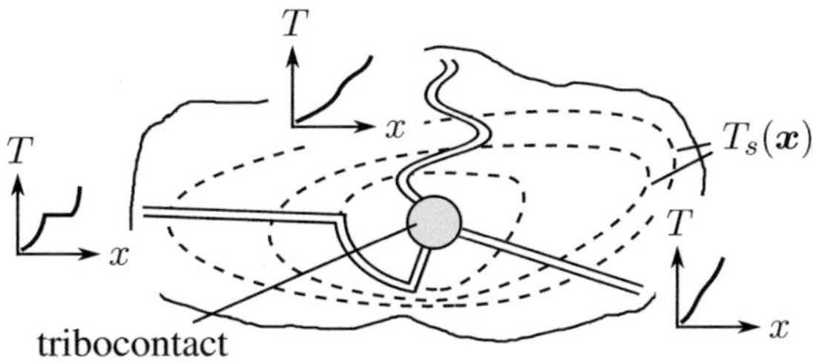

Figure 4.11.: Solid temperature distribution $T_s(\boldsymbol{x})$ determines wall temperature $T_w(x)$ along the channel

4.4.4. Viscosity influence

With temperature varying in channel direction only, the same is true for surface tension $\gamma(T(x)) = \gamma(x)$ and viscosity $\eta(T(x)) = \eta(x)$. This then directly justifies the assumption made in the derivation of equations (4.11) - (4.13). Both the Ubbelohde-Walter, eq. (2.7), and the Vogel-Cameron ansatz, eq. (2.6), for the temperature dependence of viscosity consequently permit the computation of the viscosity as a function of downstream location x.

Constant $C\,(L_u(t))$ is not any more computed with eq. (4.21) but via eq. (4.19) with details given in section 4.6. The consequences for time functions, especially the considerable speedup both with and against a thermal gradient, have already been seen in figure 4.10. It is important to remind oneself that C does not depend on x so that at any given time (i.e., for any given length of the liquid column) the flow velocity $\tilde{u}(y,z)$ at different locations along the channel is constant in x – in spite of varying viscosity values. What determines the fluid speed is a viscosity integral over the filled stretch of the channel at the time point in question, cf. eq. (4.66) and [97].

The following two figures compare fluid advancement with and against a viscosity gradient $\mathrm{d}\eta/\mathrm{d}x$ with the respective isothermal time function.

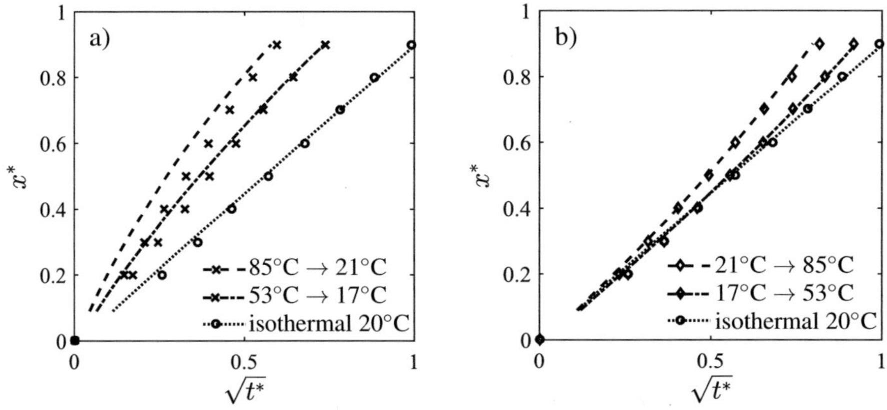

Figure 4.12.: Deviation from perfect Washburn behavior (isothermal case, straight line) of non-isothermal lubricant flow with negative (left) and positive (right) temperature gradient $\mathrm{d}T_t/\mathrm{d}x$

In the isothermal case, the time law is a straight line (if the advancement distance is plotted against the square root of time). Both theoretically and in conducted experiments this indicates classical Washburn behavior. Non-isothermal flows are faster because of the longer channel section lying in an area with solid temperature above ambient temperature as compared to the section where solid temperature is lower or around air temperature.

The capillary flow starting in the hot area experiences high velocity early while converging towards Washburn behavior in the later stages of the flow. When starting in the cold region, the behavior is vice versa i.e., approximative Washburn-type flow at the beginning which accelerates (relative to the isothermal case) towards the hotter region. For the supply of a tribocontact with lubrication fluid, this is principally favorable. Even more, figure 4.12 already includes the Marangoni effect both in experiment and in computation. Especially the flow in 4.12b) is comparatively fast even though it has to overcome the Marangoni effect which acts opposite to flow direction. Under such (realistic) conditions the tribocontact oil supply (measured in flow rate towards contact) is even better than under isothermal conditions! The reason is simple: A temperature increase lowers viscosity and thus flow resistance decreases significantly, while surface tension and contact angle reduce the capillary force to a much lesser degree. Still, the Marangoni effect may not be neglected as it plays a decisive role under conditions discussed in the following.

4.4.5. Presence and consequences of the Marangoni effect

The Marangoni effect may be relevant under non-isothermal conditions at the liquid-air boundary. Both the upper meniscus and the front meniscus are possibly concerned. As temperature and consequently surface tension were modeled to change only in fluid flow direction yet not perpendicular to the flow, and under the assumption that the extent of both meniscus and corner flow is negligible in x-direction, cf. section 4.3.1, the upper channel surface in region Ω_u is the only location where a Marangoni effect may be relevant.

Figure 4.12 suggests that the Marangoni effect plays a subordinate role to temperature induced viscosity differences. As will be seen below, this is especially true as long as temperature gradients are not strong enough to compete with or overcome the capillary force. Especially in close vicinity of the tribocontact a steeper gradient than $\mathrm{d}T/\mathrm{d}x \approx$ 10 K/mm which was realized on the described experimental equipment, cf. figure 3.11a), may be expected – based on the temperature within the contact: Both thermal elastohydrodynamic (TEHD) computation [9, 176] and easily observable temperature-induced material changes (e.g., melting and consequent deformation of engineering plastics which is exploited in section 5.2.4's experiments) allow to at least approximate a temperature field around the contact. Favorably, gradients within the contact which due to flash temperatures, cf. [157], may well be higher need no consideration due to the absence of a liquid-air interface in the (fully-filled) contact.

Besides the temperature gradient, the flow velocity u is the other important factor which influences the relative importance of the Marangoni effect compared to viscosity: The slower the flow, the more decisive is the Marangoni effect. From eq. (4.19), this may be the case in relatively long channels. The influence of the lubrication liquid's base viscosity η_n on flow velocity is not as straightforward to judge, because it has effects both in the bulk of the flow as well as at its boundary via eq. (4.19) and (4.23), respectively. Together with the fact that the Marangoni number may reach values $\mathrm{Ma} \gg 1$, this justifies more in-depth analysis.

Marangoni modeling

Even without the simplifying assumption that the microchannel in consideration has a constant oil-filled cross-section $\tilde{A}_S$, the slope of the upper interface in flow direction is negligibly small $\mathrm{d}H/\mathrm{d}x \approx 0$, cf. section 4.3.1. Then the surface normal at any point on the upper interface $\boldsymbol{n}$ can be regarded as perpendicular to x as seen in figure 4.8. This simplifies the general interfacial stress balance, eq. (2.58), considerably. From the definitions in section 2.4.4 $\boldsymbol{t}_2$ is always aligned with x whereas $\boldsymbol{t}_1$ lies in the microchannel's cross-section. Then with eq. (4.51)

$$\nabla_S T(\boldsymbol{x}) = \left.\frac{\mathrm{d}T(x)}{\mathrm{d}x}\right|_{z=H} = \frac{\mathrm{d}T(x)}{\mathrm{d}x} \tag{4.52}$$

and the tangential stress balance, eq. (2.59) reduces via eq. (2.65) to $[(\boldsymbol{\tau}_B - \boldsymbol{\tau}_A)\cdot\boldsymbol{n}]\cdot \boldsymbol{t}_1 = \tau_{M_1} = 0$ and

$$\tau_{M_2}(x) = \tau_M(x) = \frac{\mathrm{d}\gamma}{\mathrm{d}T}\frac{\mathrm{d}T(x)}{\mathrm{d}x} \tag{4.53}$$

Inserted into eq. (4.23) this defines the open channel boundary condition including the classical "no stress" case in absence of a temperature gradient. Surface tension's temperature function is in good approximation a straight line, cf. figure 3.11b), so that constant $\mathrm{d}\gamma/\mathrm{d}T$ leads to a Marangoni stress which is proportional to the local temperature gradient. With temperature modeled to be equal in the fluid and in the solid, $\tau_M \propto \mathrm{d}T_w/\mathrm{d}x$. The most critical location for a failure of the lubricant supply is consequently the point where the wall temperature gradient is highest – expectedly in close vicinity of the tribocontact.

The Marangoni force is then

$$F_M(L_u) = \int_0^{L_u} \tau_M(x) s_o(x)\mathrm{d}x \tag{4.54}$$

A major drawback of such a modeling is that it cannot cope directly with $\mathrm{d}^2T_w/\mathrm{d}x^2 \neq 0$ or equivalently nonconstant τ_M along the channel as this conflicts with the unidirectional flow assumption (see section 5.2.3 for more details regarding a similar problem). Uncritical for cross-sectional analysis at a specific location X along the channel, a workaround for fluid advancement computation is proposed in the form of an averaged, constant Marangoni stress, which is the basis for the following computations:

$$\tilde{\tau}_M := \frac{\int_0^{L_u} \tau_M(x) s_o(x) \mathrm{d}x}{\tilde{s}_o L_u} \tag{4.55}$$

Consequences of the Marangoni effect

The magnitude of the Marangoni force is a function of the size of the upper surface and therefore mainly depends on the length of the fluid column L_u. Just like the viscous force but unlike the capillary force, eq. (4.44), it grows with oil advancement. Consequently, in the very probable case of a non-decreasing temperature gradient on the way towards the tribocontact $\mathrm{d}^2T_w/\mathrm{d}x^2 \geq 0$, there exists a critical channel length where the Marangoni force balances the capillary force. This is yet not synonymous with a motionless lubricant fluid even in a quasi-stationary model. The Marangoni effect induces a Couette flow whereas capillary flow is of Poiseuille type and the resulting flow profile is a superposition of both. Figure 4.13 shows various velocity profiles in the channel center plane at $y^* = 0$:

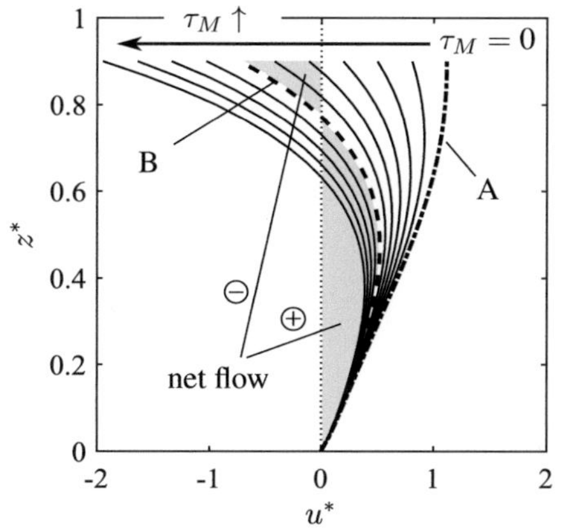

Figure 4.13.: Influence of the Marangoni effect on the flow rate: Velocity profiles in the channel center plane at $y^* = 0$ under increasing Marangoni effect due to growing $|\mathrm{d}T/\mathrm{d}x|$: see figure 4.14 for the cross-sectional velocity field $u^*(y, z)$ of cases A and B

The rightmost curve for $\tau_M = 0$ is pure Poiseuille flow with the deviation from the expected parabolic shape coming from the Gaussian channel geometry. With a growing Marangoni effect the flow velocity and with it the flow rate decline as compared to the isothermal case and the capillary can deliver less lubricant per unit time. Affected are especially near-surface fluid layers. Under a steeper local temperature gradient the velocity becomes negative here: Oil is pulled away from the tribocontact. This is less critical as it sounds as long as the fluid flow towards the tribocontact, which simultaneously takes place in deeper fluid layers, outplays the negative near-surface flow. Both can be seen as well in figure 4.14 which contrasts the velocity profiles in an open channel a) without Marangoni effect i.e., with the classical "no stress" boundary condition and b) with a Marangoni stress opposing the capillary flow as in a typical environment in the vicinity of the tribocontact.

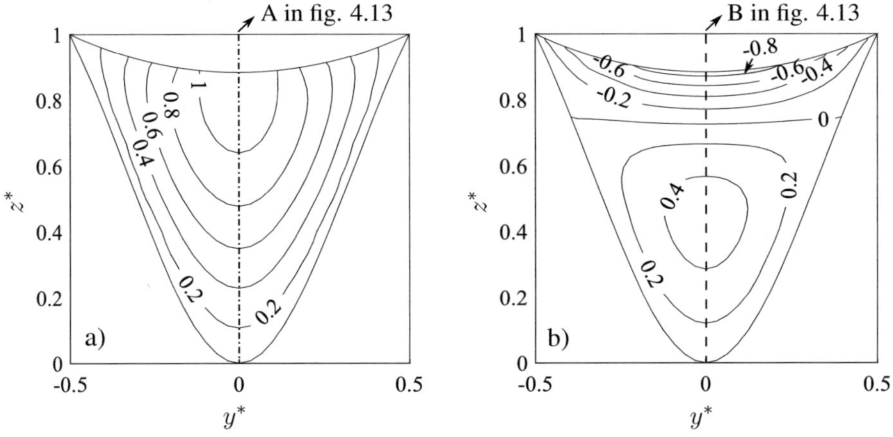

Figure 4.14.: Contour plot of the flow velocity u^*: a) base case of isothermal flow; b) the Marangoni effect pulls lubricant near the open surface away from the tribocontact which leads to simultaneous forward and backward flow in different fluid layers

Once the absolute flow rate q falls below a critical value q_{cr} needed to sufficiently supply the tribocontact, starvation occurs with all the negative effects it implies. It is not within the scope of this work to determine q_{cr}. Rather it is expected to be a result of EHD theory applied to the tribocontact in consideration (see chapter five for some more details). A definite lower bound for q_{cr} is the situation where the net flow $q = 0$. The sum of the shaded areas in 4.13 illustrates the concept for the fluid slice in the channel center plane which has to be extended to the whole cross-section. Net flow $q = 0$ does not necessarily mean that the lubricant flow is at rest. In analogy to figure 4.14 a circling flow appears which flows towards tribocontact at the channel bottom and back in the upper fluid layers, yet without supplying the tribocontact [118]. For even steeper temperature gradients the net flow becomes negative – a situation to be avoided.

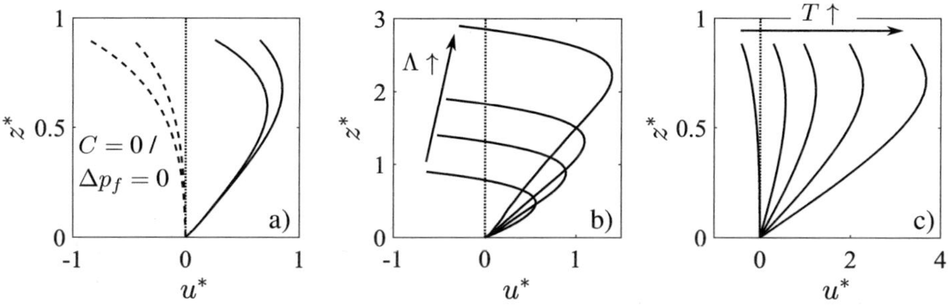

Figure 4.15.: Velocity profiles at $y^* = 0$ (Gaussian channel) show that the consequences of the Marangoni effect can be influenced a) by the neglection of the capillary pressure (dashed lines) compared to the regular case (solid lines), b) by the aspect ratio, and c) by the absolute temperature; computation of b) based on Lenormand's formula for a rectangular channel (see text)

The great advantage of capillarity driven channel flow is seen in figure 4.15a), solid lines: It is able to support the Marangoni effect to a considerable degree while still maintaining a positive net flow towards the tribocontact. Without the pressure drop at the capillary front meniscus any positive temperature gradient (even a minor one) leads to negative flow in all fluid layers (see dashed lines). This is the case for a thin lubricant film on a flat solid body – probably the standard situation in tribological applications today. Certainly, surface roughness may produce some degree of capillary force, which is yet small compared to a microchannel's, cf. section 3.1.

A strategy to further minimize the Marangoni influence for a given desired flow rate is to increase the aspect ratio, as shown in figure 4.15b): The Marangoni force is proportional to the upper surface area A_u^I and therefore via $s_o(x)$ is affected by the channel width. Deeper channels come with the advantage of allowing more throughput without increasing the surface area. First experiments conducted and presented in [142] were able to demonstrate the reduction in Marangoni influence although further experimental refinement is needed for a higher reliability of the results. To be precise from a modeling perspective, the dependence of G on the aspect ratio has not been thoroughly investigated, so that for the computation in the figure, Lenormand's factor $K_{Len}^{(\square)}$, eq. (2.78), has been applied even though it was developed exclusively for rectangular channel cross-sections. While figure 2.11 suggests that for aspect ratios $1 \leq \Lambda \leq 3$ it has a negligible influence, a rigorous analysis for elliptical and Gaussian cross-sections is pending.

Critical are also low-temperature environments: Figure 4.15c) shows changes in channel center plane flow profile due to different levels of absolute temperature, yet with constant temperature gradient. From the latter, the Marangoni effect is constant, yet high viscosity at low temperature leads to a slow flow velocity and consequently the Marangoni effect becomes relatively more important. Mathematically, this case tends towards the absence of the front meniscus pressure drop discussed above and depicted in figure 4.15a). Given a constant Marangoni stress the Couette contribution to total flow is constant as well. What changes is the Poiseuille part, expressed in the driving factor C. Eq. (4.66) will show how the pressure drop and viscosity combine with lubricant column length and inertial forces from a non-inertial frame of reference to determine C.

4.5. Moving structured bodies

In a tribocontact there is, per definition, relative motion. At least in some cases one of the two contacting bodies can be assumed fixed in an inertial frame of reference. Yet if this is the case, the other one must be moving relative to this frame. The analysis of non-inertial frames is therefore indispensable.

Two exceptions must be mentioned: First, capillary supply structures do not need to be fabricated in both bodies, and therefore the body in the inertial frame could be chosen. This yet restricts design flexibility. Furthermore, various counterexamples with both bodies moving exist, in gears for example. Secondly, the motion and its consequences may be small enough so that they can be neglected and a moving frame can still be treated like an inertial frame. Here, an appropriate decision basis is required to determine when such an assumption is justified. It is given in the following, based on further nondimensionalization.

Movement may not only arise in the tribocontact itself, but also the object which contains the tribocontact may move relative to an inertial frame e.g., a subassembly in rotating machinery as well as a vehicle. Also vibration can be the cause of significant accelerations, both in rotation and in translation.

That such movement can significantly influence the fluid flow in the vicinity of the tribocontact has been observed e.g., by Liang et al. [131] in centrifugation experiments. In contrast to tribological applications, research of thin film and microchannel flow in non-inertial reference frames concentrates on the fact how to purposefully control a rotation to make a fluid move in the desired way. Examples are the "spin coating" process of disks or "Lab-on-CD" applications, cf. sections 1.2 and 2.3.4. This work yet studies capillary force as the driver of lubricant liquid flow which has to cope with the structured part's acceleration as an externally given disturbance, so that angular velocity, translational, and/or angular accelerations are analyzed with the focus on how they might actually hinder the lubricant flow. Certainly, the inertial forces in a moving frame may as well support the capillary force in driving the flow rather than being opposed to it. Yet only the latter is critical for design. Consequently, the following discussion has the character of a worst case analysis.

4.5.1. Significance of inertial forces from structure movement

Centrifugal, Euler, and translational acceleration contributions

The variables controlling movement are nondimensionalized as follows:

$$\boldsymbol{\omega}^* := \frac{\boldsymbol{\omega}}{\omega_0} \tag{4.56}$$

$$\boldsymbol{\alpha}^* := \frac{1}{\alpha_0}\frac{\mathrm{d}\boldsymbol{\omega}}{\mathrm{d}t} = \frac{\boldsymbol{\alpha}}{\alpha_0} \tag{4.57}$$

$$\boldsymbol{a}_T^* := \frac{1}{a_{0T}}\frac{{}^I\mathrm{d}^2}{\mathrm{d}t^2}\breve{\boldsymbol{r}} = \frac{\boldsymbol{a}_T}{a_{0T}} \tag{4.58}$$

The characteristic parameters $\omega_0, \alpha_0, a_{0T}$ all are defined as independent ones instead of being interconnected by a characteristic time scale. The latter would basically require the definition of a second characteristic time scale as the previously introduced dimensionless time which relates fluid velocity and channel length is only coincidentally appropriate for part accelerations. These may for example be governed by characteristic times which are shorter by various orders of magnitude. If such time dependent behavior is of interest e.g., in vibrational analysis, further complexity arises from the fact that a cornerstone of the model so far, namely the complete neglection of inertia terms (by which the left hand side of eq. (4.1) is meant here) must be strongly put into question. If and how existing theory in oscillations of a slender channel, especially with flexible walls [6, 7], can sensibly be harmonized with the present model remains an open question. As will be seen, the simplifications made above still allow to characterize the flow behavior with respect to critical solid body accelerations and critical angular velocity and therefore enable the tribologist to judge if certain capillaries can maintain the supply of the tribocontact with lubricant.

With the above definitions and with equations (4.9) and (4.10) the inertial force densities due to centrifugal, Euler and translational accelerations are nondimensionalized as shown in table 4.6.

Table 4.6.: Nondimensionalized force densities and dimensionless factors $\varkappa_j$ in non-inertial reference frames

centrifugal	$\boldsymbol{f}_Z = -\rho\,\boldsymbol{\omega} \times (\boldsymbol{\omega} \times \breve{\boldsymbol{q}}) = \underbrace{\rho \omega_0^2 \breve{q}_0}_{f_{0Z}} \left(\underbrace{-\boldsymbol{\omega}^* \times (\boldsymbol{\omega}^* \times \breve{\boldsymbol{q}}^*)}_{\boldsymbol{f}^*_Z} \right)$ $\xi f_{0Z} = \varkappa_Z = \delta^2 \frac{\rho \omega_0^2 l_0^2 \breve{q}_0}{\eta_0 u_0} = \frac{\rho \omega_0^2 h_0^2 \breve{q}_0}{\eta_0 u_0} = \delta^2 \mathrm{Re} \frac{\omega_0^2 \breve{q}_0 l_0}{u_0^2}$
Euler	$\boldsymbol{f}_E = -\rho\,\frac{\mathrm{d}\boldsymbol{\omega}}{\mathrm{d}t} \times \breve{\boldsymbol{q}} = \underbrace{-\rho \alpha_0 \breve{q}_0}_{f_{0E}} \left(\underbrace{\frac{\mathrm{d}\boldsymbol{\omega}^*}{\mathrm{d}t^*} \times \breve{\boldsymbol{q}}^*}_{\boldsymbol{f}^*_E} \right)$ $\xi f_{0E} = \varkappa_E = \delta^2 \frac{\rho \alpha_0 l_0^2 \breve{q}_0}{\eta_0 u_0} = \frac{\rho \alpha_0 h_0^2 \breve{q}_0}{\eta_0 u_0} = \delta^2 \mathrm{Re} \frac{\alpha_0 l_0 \breve{q}_0}{u_0^2}$
translat. acc.	$\boldsymbol{f}_T = -\rho\,\frac{{}^I\mathrm{d}^2}{\mathrm{d}t^2} \breve{\boldsymbol{r}} = \underbrace{-\rho a_{0T}}_{f_{0T}} \left(\underbrace{\frac{{}^I\mathrm{d}^2 \breve{\boldsymbol{r}}^*}{\mathrm{d}t^{*2}}}_{\boldsymbol{f}^*_T} \right)$ $\xi f_{0T} = \varkappa_T = \delta^2 \frac{\rho a_{0T} l_0^2}{\eta_0 u_0} = \frac{\rho a_{0T} h_0^2}{\eta_0 u_0} = \delta^2 \mathrm{Re} \frac{a_{0T} l_0}{u_0^2}$

Here, $\varkappa_j$ with $j = \{Z, E, T\}$ are dimensionless numbers judging the relative influence of centrifugal and Euler force densities as well as translational acceleration compared to pressure gradients and viscous contributions in the derived governing equation set (4.11) - (4.13): If on the one hand $\varkappa_j \ll 1$, especially if $\varkappa_j \lessapprox \delta^2$ the corresponding effect is insignificant in magnitude compared to pressure and/or viscous forces, and the non-inertial frame of reference can be treated as if it was an inertial one. From eq. (4.12) and (4.13) the force densities may even be greater in lateral and vertical directions by a factor of δ to still be negligible. If on the other hand $\varkappa_j \gg 1$, the respective force density clearly dominates the flow. In such a case a change in design is required, because the capillary force is simply not strong enough to maintain the lubricant supply of the tribocontact (given relevant force directions as discussed in the following section). Both situations can be understood as approximate bounds.

Only in between the two bounds i.e., for

$$\frac{1}{l_0}\sqrt{\frac{\eta_0 u_0}{\rho \breve{q}_0}} \lessapprox \omega_0 \lessapprox \frac{1}{h_0}\sqrt{\frac{\eta_0 u_0}{\rho \breve{q}_0}} \tag{4.59}$$

$$\frac{1}{l_0^2}\frac{\eta_0 u_0}{\rho \breve{q}_0} \lessapprox \alpha_0 \lessapprox \frac{1}{h_0^2}\frac{\eta_0 u_0}{\rho \breve{q}_0} \tag{4.60}$$

$$\frac{1}{l_0^2}\frac{\eta_0 u_0}{\rho} \lessapprox a_{0T} \lessapprox \frac{1}{h_0^2}\frac{\eta_0 u_0}{\rho} \tag{4.61}$$

a more detailed analysis of the fluid flow, based on the derived equation set, is required in order to judge the remaining capillary transport capability in terms of flow rate q.

Coriolis force

Coriolis force density has been excluded up to now as it is the only force density which does not only influence the flow velocity but is itself dependent on it. Treatment as above leads to:

$$\xi f_{0C} = \varkappa_C = \delta^2 \frac{2\rho\omega_0 l_0^2}{\eta_0} = \frac{2\rho\omega_0 h_0^2}{\eta_0} = \delta^2 \mathrm{Re}\frac{2\omega_0 l_0}{u_0} \tag{4.62}$$

and

$$\boldsymbol{f}_C = -2\rho\,\boldsymbol{\omega} \times {}_O\boldsymbol{u} = -2\rho\omega_0 u_0 \begin{pmatrix} \delta\left(\omega_y^* w^* - \omega_z^* v^*\right) \\ \omega_z^* u^* - \delta\omega_x^* w^* \\ \delta\omega_x^* v^* - \omega_y^* u^* \end{pmatrix} \approx -2\rho\omega_0 u_0 \begin{pmatrix} 0 \\ \omega_z^* u^* \\ -\omega_y^* u^* \end{pmatrix} \tag{4.63}$$

The last approximation step is justified by two reasons: First, v and w have already been neglected in section 4.1 in the context of unidirectional flow. Even if this had not been the case the long and slender nature of the channel now suggests the same via the neglection of terms including δ. Consequently, a relevant Coriolis force influence is expected only in the coordinate directions perpendicular to the main flow direction u. This is an obvious result considering that the Coriolis force is proportional to the flow velocity and lateral and vertical velocities are minimal. The fact that y- and z-directions in eq. (4.12) and (4.13) involve an additional small factor δ compared to eq. (4.11) which further reduces the influence of the Coriolis force density only contributes to the above stated and underlines the challenges a Coriolis force based research faces for example in fluid mixing applications, which need high downstream velocity to be able to generate some crossflows at all [130, 181]. In conclusion, the Coriolis force is least likely to influence the capillary transport in this research context.

4.5.2. Integration into the existing equation framework

The nondimensionalization above led to approximate bounds as the first prerequisite for inertial force densities to be significant yet did not consider the effective direction of the respective forces. Body forces have been carried through so far only for the sake of simplicity. Only in special cases electromagnetic forces may be relevant in a tribological context and gravity can generally be neglected due to a small Bond number $\mathrm{Bo} \ll 1$, cf. section 2.1. If the Coriolis force influence is equally neglected with the above reasoning, the force density vector (2.44) based on (2.43) simplifies to:

$$\boldsymbol{f} = \boldsymbol{f}_f = \rho \left(\boldsymbol{\omega} \times (\boldsymbol{\omega} \times \breve{\boldsymbol{q}}) + \frac{\mathrm{d}\boldsymbol{\omega}}{\mathrm{d}t} \times \breve{\boldsymbol{q}} + \frac{{}^I\mathrm{d}^2\breve{\boldsymbol{r}}}{\mathrm{d}t^2} \right) \tag{4.64}$$

Due to the absence of both nonlinearity and flow velocity influence this equation is mathematically extremely favorable: The superposition of effects allows to externally predetermine and combine the different force influences resulting from structure movement which then enter the governing lubricant liquid flow equations. The force density from a translational acceleration of a certain point on the rigid body is not affected by an advancing capillary flow.

The only variable which is not independent of the fluid's advancement in the channel is $\breve{\boldsymbol{q}}(\boldsymbol{x})$. If $|\breve{\boldsymbol{q}}| \gg y_0, z_0$, especially if $\breve{\boldsymbol{q}}$ is comparable or bigger than the characteristic channel length x_0, the dependence can further be reduced in analogy to the basic long and slender channel assumption $x_0 \gg y_0, z_0$, eq. (4.2), to result in:

$$\breve{\boldsymbol{q}}(\boldsymbol{x}) \approx \breve{\boldsymbol{q}}(x) \, \forall \, y, z \tag{4.65}$$

This approximation fulfills the validity condition stated in the derivation of eq. (4.14) and thereby of the governing microchannel flow equation (4.18), namely that the force density component in channel direction f_x is a function of x only. This has two consequences: First, unlike in typical non-inertial microfluidics models as described in section 2.3.4 the determination of the fluid column's center of gravity is not necessary at this point (with more details given in section 4.6). Secondly, from eq. (4.14) and (4.19) $\partial p/\partial x$ is a function of x only. Consequently, the pressure distribution $p(y, z)$ in every cross-section along the channel is similar. For this to be true, eq. (4.12) and (4.13) require f_y and f_z to be independent of x – as later implemented in eq. (4.15) and (4.16). This conflicts with the nature of at least some inertial forces e.g., the force due to an Euler acceleration along a radially oriented channel on a spinning disk. To cover these forces, a general three-dimensional pressure field is necessary which allows for crossflows v and w to balance such inertial contributions – similar to what is found e.g., in [130]. These crossflows cannot be explained by this work's unidirectional flow model which is consequently not capable of handling any arbitrary $f_y(x)$ and $f_z(x)$. Yet as the nondimensionalization procedure in section 4.1 characterized v and w as negligible, one must conclude that although x-dependent f_y and f_z may require the pressure field to adapt, this is not decisive for the overall downstream flow towards tribocontact. The findings in [168] support this for Reynolds numbers which are typical in capillary flow context.

4.5.3. Consequences for capillary flow

Focusing on downstream direction, the resulting Poisson equation (4.18) which governs the unidirectional microchannel flow reveals a very simple relationship between the downstream pressure gradient created by the capillary front meniscus and the force density in x-direction: If $\partial p_m/\partial x - f_x(x) < 0$ i.e., if the pressure gradient is stronger in absolute value than the combined inertial force density, the capillary flow prevails over the influence of the movement of the channel itself: The capillary flow is slower than in the comparable inertial frame yet is not stopped. The latter happens in the case when $\partial p_m/\partial x = f_x$. If combined inertial forces are greater in absolute value than the pressure gradient, capillary flow cannot develop. Important to note is that with constant front meniscus pressure drop Δp_f the pressure gradient diminishes in absolute value the more the lubricant advances i.e., the longer the fluid column extends in the channel. In the presence of a counteracting force – in the simplest case a constant one – there is forward flow initially which stops once the pressure gradient and the force balance.

Figure 4.16 shows results from the basic setup of a rotating disk with radial capillary flow towards the rotation axis. By controlling the angular velocity the fluid flow can be stopped and reversed (an example is the solid line case). A constant angular velocity (dashed line) leads to an equilibrium state. The minor fluctuations around it have numerical reasons although physical effects not covered by the model might in reality also be present. Experiments to validate the computation which have not been conducted in this work might be adapted from the design of Leung and Ren [129]. Different from the experimental setup used here, cf. section 3.3, they use a high-speed camera which rotates with a spinning disk.

The effect which the solid body movement has is cumulative with the Marangoni effect: From above the inertial forces and the pressure gradient combine to determine the strength of the Poiseuille flow component measured in factor $C(L_u(t))$, which will be determined in eq. (4.66). The Marangoni effect then superposes a Couette flow contribution: Qualitatively not different from what was discussed in section 4.4.5, it yet leads to a much earlier breakdown of the tribocontact's lubricant supply if inertial forces from the capillary structure's movement already weaken the Poiseuille flow.

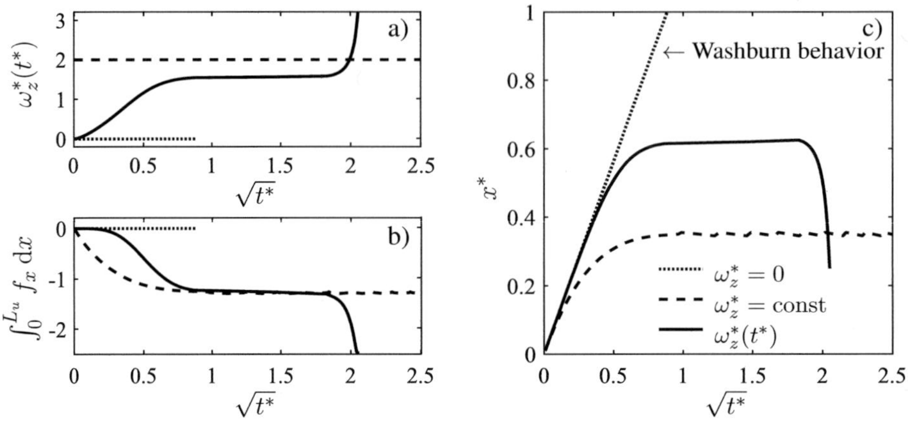

Figure 4.16.: Capillary channel flow in a rotating reference frame with flow direction radially towards the rotation axis ("spinning disk" case): no rotation (dotted line), constant angular velocity (dashed line) and varying angular velocity (solid line) as specified in a)

In conclusion, any inertial force resulting from a non-inertial reference frame (with the exception of Coriolis force) cannot only hinder but suppress capillary flows, irrespective if it is of centrifugal or Euler nature or if it is the consequence of a translational acceleration. This is true as long as the force component in channel direction is strong enough to overcome the prevailing pressure gradient. As the latter gets weaker with lubricant advancement, care has to be taken that the flow does not stop before it reaches the tribocontact, especially in longer channels and even more with a thermal gradient present. The effect the structured body's movement has on the pressure gradient itself is discussed in what follows.

4.6. Pressure considerations

All computational results presented so far implied a known flow driving factor C. It is determined by applying pressure boundary conditions at the start and the end of the liquid lubricant column. Pressure at the channel inlet has been defined so far to equal ambient conditions, $p(x{=}0) = p_{am}$. Pressure at the end of the fluid column, right inside the front meniscus, is a result from front meniscus pressure jump $p(x{=}L_u) = (p_{am} - \Delta p_f)$. Due to the assumptions made in section 4.3.1, both conditions are the classical microflow pressure boundary conditions. Applying them after integration over the total fluid column length to eq. (4.19) yields:

$$C(L_u) = \left(-\Delta p_f - \int_0^{L_u} f_x(x)\,\mathrm{d}x \right) \frac{1}{\int_0^{L_u} \eta(x)\,\mathrm{d}x} \tag{4.66}$$

Exemplary computations of the force integral are displayed in figure 4.16b). In the absence of inertial forces from structure movement and with constant viscosity the above equation simplifies to eq. (4.21). C then allows to compute the velocity profile in the channel's cross section with the help of eq. (4.20) for a given fluid column length L_u. It also reveals the character of the pressure function along the fluid column when integrating eq. (4.19), this time from 0 to x:

$$p(x) = \int_0^x f_x\,\mathrm{d}\zeta + C(L_u) \int_0^x \eta\,\mathrm{d}\zeta + p_{am} \;\forall\; x \in (0, L_u) \tag{4.67}$$

In the basic case where $f_x = 0$ and $\eta = \mathrm{const}$, pressure decreases linearly from its maximum $p_{max} = p_{am}$ at the channel inlet to its minimum $p_{min} = p_{am} - \Delta p_f$ at the front meniscus. From eq. (4.64) inertial forces from a moving reference frame obey $\mathrm{d}f_x/\mathrm{d}x = \mathrm{const}$ which leads to a parabolic pressure function. This is in line with [113, 132, 168]. In such a case, the local pressure gradient adapts to balance the inertial force density variation along the channel in order to ensure a constant C and a constant $\tilde{\bar{u}}$ which is required by continuity, eq. (4.8).

The parabolic pressure function is independent of the presence of surface tension gradients and/or an upper meniscus as modeled here. Pressure yet deviates from the parabola once viscosity cannot be treated as constant along the channel any more, for example when the temperature distribution along the channel requires Ubbelohde-Walter or Vogel-Cameron models, eq. (2.6) and (2.7). Nevertheless, the pressure function inside the fluid column is not of superior interest as it does not tell if and how well a capillary supply of tribocontact can be maintained. It is rather the absolute value of C, together with the Marangoni effect, which is decisive for the avoidance or at least minimization of starvation.

4.7. Flow rate and time function of fluid advancement

The importance of the time function of lubricant fluid advancement mainly results from experimental necessity, cf. section 3.3, as it only allows to compare model and experiment. On the basis of low Reynolds or reduced Reynolds numbers microchannel flow was argued to be dominated by viscous forces whereas inertial contributions due to the acceleration of fluid parcel masses inside the fluid column were neglected, cf. section 4.1. As a consequence, the stationary flow at a certain point in time analyzed until now can be generalized towards the time function of the flow using a quasi-stationary ansatz. Such reasoning only allowed to derive the classical Washburn equation (2.71) via eq. (2.70). The approach presented in the following is similar. Obviously, if not already stated above, flow velocity $u(t)$, flow rate $q(t)$, advancement length $L_u(t)$ and geometrical constant $C(L_u(t))$ are all treated as time functions.

Influence of flow driving factor C

At first, a basic case of an open channel with arbitrary but constant cross-sectional shape in an inertial frame of reference without additional body forces and uniform temperature field is discussed, so that the Marangoni effect does not need to be considered and viscosity is constant. In such case, $C(L_u(t))$ is governed at any given point in time by eq. (4.21). As from the assumptions in section 4.3.1 the front meniscus pressure drop is constant both in space and in time, the only time influence on $C(t)$ is $L_u(t)$. From Washburn logic it is expected that the advancement of the fluid column slows down with time i.e., the longer it gets. This is expressed in the hyperbolic relationship $C(t) \propto 1/L_u(t)$ which essentially describes the (in absolute value) diminishing "strength of flow" $C(t)$ over time. As a consequence of eq. (4.20) and the appropriate boundary conditions, downstream flow velocity u decreases over the whole channel cross-section proportional to C. The same is true for averaged velocity $\tilde{\bar{u}}(t)$ and for flow rate $q(t)$. Then,

$$q(t) \propto \frac{1}{L_u(t)} \tag{4.68}$$

$C(t)$, $\tilde{\bar{u}}(t)$ and $q(t)$ never reach zero, which means the capillary flow will theoretically never stop, even though it slows infinitely. In these characteristics such a basic case shows perfect Washburn behavior.

This does not apply in the general case where $C(t)$ obeys eq. (4.66) instead of eq. (4.21). Especially for $C(t) = 0$ flow stops or if $C(t) > 0$ it even changes direction, cf. eq. (4.20), and backward flow occurs as seen in figure 4.16. Eq. (4.66) shows that this happens once the integral of the total force density along the fluid column acting in channel direction balances or exceeds the meniscus pressure drop, respectively. In the latter case the capillary flow recedes even though the (basically) unchanged meniscus curvature suggests advancing flow. What might seem clear is that varying viscosity cannot by itself stop or reverse a capillary flow, because it is always positive per definition which is then true as well for its integral in eq. (4.66). That a Marangoni effect is capable of flow reversal has already been seen. It does not influence Poiseuille flow factor $C(t)$ because it induces a Couette flow. Its consequences can therefore be judged by analyzing the respective velocity profile as has been done in section 4.4.5, yet on the basis of the correct $C(t)$: If the driving force exerted by the capillary front meniscus has already been weakened by a counteracting force due to structured body motion for example, a comparably small Marangoni influence might be sufficient for the net flow towards tribocontact to become negative, cf. figure 4.15a). Thus even though the influence of a single effect might be bearable in a given case, their superposition can lead to a critical state irrespective of the nature of flow they cause.

Time function characterization depending on the combination of influencing effects

In analogy to the derivation of Washburn's equation in section 2.5.1, the combination of the flow rate equation (2.69) and the cross-sectional velocity integral, eq. (4.43), leads to an ODE which governs the general time behavior of capillary flow – including all effects discussed previously:

$$\frac{\mathrm{d}L_u}{\mathrm{d}t} = \frac{1}{\tilde{A}_S} \iint_{(\tilde{A}_S)} \tilde{u}\left(y, z, C\left(L_u\left(t\right)\right)\right) \,\mathrm{d}y \,\mathrm{d}z \tag{4.69}$$

For an arbitrary cross-sectional shape the evaluation of the integral suggests numerical solution techniques, cf. section 4.2.2. In the special case of an open rectangular cross-section with flat upper meniscus, the velocity profile $\tilde{u}(t)$ is given via eq. (4.28) and (4.31) as $\tilde{u}(y, z, \tau_M, C(t))$ i.e., as a function of both Marangoni stress and Poiseuille constant $C(t)$, which is analytically integrable. This allows to further simplify eq. (4.69) and reveal the characteristic time behavior under the different investigated effects. Table 4.7 lists the resulting differential equations for six different cases A-F. Various constants, especially those from the underlying Fourier series approach, have been concentrated into c_j.

Case A, capillary transport without additional effects leads to classical Washburn behavior, $L_u(t) \propto \sqrt{t}$ with given initial condition $L(t{=}0) = 0$ which is also assumed in all following cases. The fluid speed is governed by constant c_1 which includes both the strength of the front meniscus and the fluid resistance due to the solid channel boundaries just like in eq. (2.71). Surface stress resulting from the Marangoni effect due to thermal gradients is present in case B. It leads, in absence of a capillary pressure, to a linear time function. The fluid velocity is proportional to the magnitude of the thermal gradient. Even though the total viscous force increases with fluid column length, this pure Couette flow does not slow down because the upper surface where τ_M acts increases likewise. Pure structured body motion with constant angular velocity $\boldsymbol{\omega}$ and translational acceleration $\boldsymbol{a}_T$, case C, provokes an exponential flow advancement function. Combining it with a capillary force in case D results in a Chini type differential equation. The assumption of constant $\boldsymbol{a}_T$ and $\boldsymbol{\omega}$ as in figure 4.16 (dashed line) may be sufficient for a worst case analysis; if both are allowed to be functions of time, $c_5(t)$ and $c_6(t)$ are time-dependent and can thus widely influence the resulting time law with an example given in figure 4.16 (solid line).

Table 4.7.: Differential equations governing the time law of fluid advancement in different cases; valid for an open rectangular channel cross-section with flat upper meniscus with $\boldsymbol{\omega}$ and $\boldsymbol{a}$ constant and initial condition $L_u(t{=}0) = 0$; the equations also serve to characterize flow rate q

case		DE: $\frac{\mathrm{d}L_u}{\mathrm{d}t} = \dots$	time law
A	capillary pressure only $\tau_M = 0, \eta = \text{const}, f_x = 0$	$c_1 \frac{1}{L_u}$	square root (Washburn): $L_u(t) = \sqrt{2c_1 t}$
B	Marangoni effect only $\Delta p = 0, \eta = \text{const}, f_x = 0$	$c_2 \tau_M$	linear: $L_u(t) = c_2 \tau_M t$
C	moving structure only $\Delta p = 0, \tau_M = 0, \eta = \text{const}$	$c_3 L_u + c_4$	exponential: $L_u(t) = (c_4/c_3)(\exp(c_3 t) - 1)$
D	cap. pressure, moving structure $\tau_M = 0, \eta = \text{const}$	$c_5 L_u + c_6 + c_7 \frac{1}{L_u}$	solution to Chini eq.
E	cap. pressure, nonconstant visc. $\tau_M = 0, f_x = 0$	$c_8 \frac{1}{\int_0^{L_u} \eta(x)\mathrm{d}x}$	
F	all effects considered	$c_9 L_u + c_{10}\tau_M + c_{11} + c_{12}\frac{1}{\int_0^{L_u} \eta(x)\mathrm{d}x}$	
		$\frac{q}{\tilde{A}_S} = \dots$	

Analytical insight is limited in the case of variable viscosity (E) even without structure movement, cf. figures 4.10 and 4.12, due to the nature of empirical Vogel-Cameron or Ubbelohde-Walter laws, eq. (2.6) and (2.7). Considering the fact that this section's analytical analysis was based on a rectangular channel cross-section without upper meniscus allowed, numerical treatment is inevitable. The procedure is straightforward: First, a vector of discrete distances $L_u(i)$, $i = 1..N$ with $L_u(i = 1) = 0$ and $L_u(i = n{+}1) - L_u(i = n) \ll l_0$ is defined. Starting an iterative scheme with $n = 2$, $C(L_u(i{=}n))$ is computed on the basis of eq. (4.66), the lubricant velocity is solved for with eq. (4.27) which then determines the time for the flow to reach the next grid point, namely $L_u(i{=}n{+}1)$. Like this, a time vector is constructed. The computation domain is not predetermined but is identical to the lubricant column length at any given time. As a consequence, the location of the driving lubricant-air interface does not need to be computed (in contrast to e.g., phase field or level set approaches) and its characteristic pressure difference can always be treated as a boundary condition which notably contributes to computational speed. In fact, all computations shown in this work have been executed on a laptop computer (2.7GHz quad-core, 16 GB RAM) in few minutes or less time.

Although important for experimental validation, the time function of fluid advancement is – from a tribologist's view – less of interest compared to flow rate q which is vital in tribocontact supply. As $q(t) = \tilde{A}_S \cdot (\mathrm{d}L_u/\mathrm{d}t)$, eq. (2.69), the third column in table 4.7 equally informs about flow rate dependence e.g., on fluid column length. A simple guideline for microchannel design, case A illustrates the inversely proportional relationship between q and L_u in the isothermal case. Given tribocontact demand in lubricant volume per unit time, the maximum possible capillary length is simply determined by c_1. For a rectangular channel cross-section all factors c_j can be derived analytically. Other cross-sectional geometries require numerical computation as they do not adhere any more to the functions listed in table 4.7, although their mathematical characteristics can serve as a guideline. Major uncertainty is yet based on the simple fact that exact lubricant demand of a tribocontact is not a well-known figure, a topic raised in the following chapter.

In conclusion, the various effects may lead to very different results with respect to the timely behavior of capillary flow. Even though a capillary force alone continuously drives the lubricant liquid forward, care must be taken in order to ensure sufficient supply of the tribocontact in the presence of counteracting effects.

5. Lubricant transition into and out of a channel or surface structure

After having concentrated on liquid lubricant transport fully within a surface structure, especially within a microchannel, this chapter intends to build the bridge to adjacent topics, in particular to the tribocontact itself. Besides some preliminary research it will offer a variety of open questions which cannot be answered by only focusing on transport in a surface structure. The first section deals with the transition of oil out of a reservoir e.g., a grease and into the structured solid surface. Section 5.2 then concentrates on the transition of oil out of the capillary transport structure and into the tribocontact as the lubricant's final destination.

5.1. Oil release from a lubricant reservoir

In the basic case the lubricating liquid enters the capillary structure out of a liquid reservoir into which the structure extends. Such a reservoir may be a tank or container or – on a smaller scale – an oil drop or a surface cavity like the structured reservoirs used for experiments as depicted in figure 3.7. If these reservoirs are sufficiently large compared to the channels, especially if their volume is larger than the capillaries' volume by orders of magnitude so that their pressure level during channel filling stays approximately constant and velocities inside the reservoir are low, then it can be assumed that they do not interfere in a significant way with the flow in the channels and are able to supply the fluid amounts which the surface structure is capable of transporting. As a consequence, it is sufficient to analyze the channel flow itself.

This is not necessarily true any more if the available amount of lubricating liquid is still large compared to the channels' volume yet bound in a restricting medium – as in the case of lubricating oil bound by a thickener structure within a grease, cf. figure 2.1. For the wide application of greases in lubrication, this is dealt with in some more detail:

Oil transition from a grease into a capillary structure

While influences of temperature, pressure and inertial forces on oil release of a grease have been investigated (see section 2.2.1), the exact physics of oil bleeding under isothermal and isobaric conditions is still unknown. Lugt [136] states the probable influence of capillary forces and proposes to include them into the model of Baart et al. [5], cf. figure 2.1. In such a case, both the effect of capillary menisci in the empty space between thickener fibers and the menisci along fibers (cf. figure 1.3 for the classical understanding of liquids on fibers) have to be simultaneously considered. Combined with Baart's model this results in what is depicted in figure 5.1a).

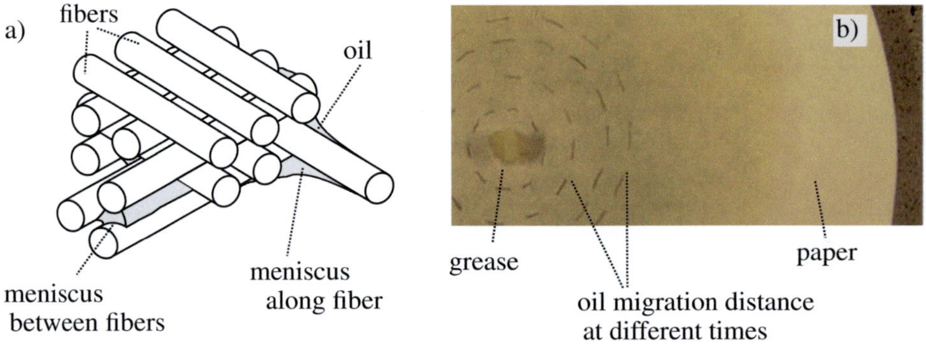

Figure 5.1.: a) Expected capillary effects within a grease if the thickener structure is modeled as proposed by [5]; b) interplay of two capillary systems: grease and carrier structure (pictured is the quantitative ashless cellulose filter paper grade 589/2, Whatman brand, GE Healthcare, USA)

Given a defined volume of oil in the grease in consideration, an equilibrium of capillary (and maybe gravitational) forces is expected – with pressure inside the oil possibly not at ambient pressure level so that static curved menisci may appear in line with the equilibrium contact angle requirement (2.15) and Young-Laplace equation (2.63). This still does not explain the release of oil out of the thickener structure.

If the grease, seen as a capillary system, is yet brought into contact with another capillary system e.g., a body with designed capillary surface structures or simply a rough surface, then the above equilibrium may be disturbed and capillarity induced oil bleeding out of the grease may occur.

Preliminary transition experiments

Preliminary experiments have been conducted with grease drops on a polished metal, a rough metal surface and on a surface with microchannels. While the latter two clearly were able to remove oil from the grease, this was not observed with the polished surface. In comparison to the previous chapters' experiments which lasted several seconds, oil bleeding out of the grease is significantly slower. Therefore, additional tests were conducted with various types of (filter) paper as carrier structures, cf. figure 5.1b), which allowed to better cover a characteristic oil bleeding time scale in the range of hours or days: The moving oil front inside the paper was visually detectable and marked on the paper at different times.

The resulting time functions of oil advancement did not directly show a square root relationship of distance and time as possibly expected from Washburn's (2.71) and Darcy's (2.53) equations, cf. figure 5.2a). The behavior is still characteristic as transport is not one-dimensional but happens in the two-dimensional paper plane. Indeed, the oil soaked area was found to be proportional to the square root of time which is equivalent to the proportionality of radial distance to the fourth square root of time as depicted in figure 5.2b).

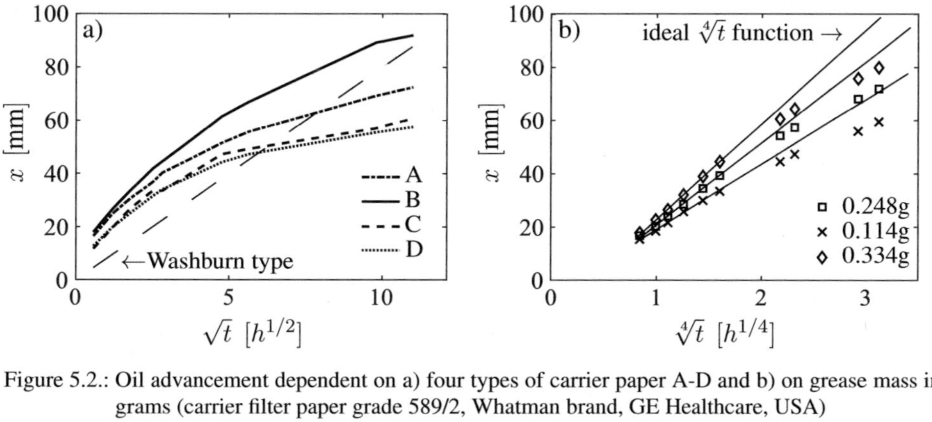

Figure 5.2.: Oil advancement dependent on a) four types of carrier paper A-D and b) on grease mass in grams (carrier filter paper grade 589/2, Whatman brand, GE Healthcare, USA)

Figure 5.2a) shows the migration behavior on four different types of carrier paper A-D with different porosity, expressed in the respective proportionality constant κ_D: Oil bleeding speed is very well dependent on the structure, the grease is placed upon. While the results of section 3.3's experiments of microchannel oil advancement expectedly proved independent of the total amount of oil in the reservoir, figure 5.2b) indicates that a larger grease drop leads to faster oil advancement. This is at least in part due to the different contact areas between grease drops and substrate: A comparison of a flat grease drop with large contact area showed slightly faster transport than an equal-weight slim and high grease drop with small contact area. While in the beginning all experimental data agree very well with the ideal fourth root function, towards larger times a general slowdown is observed as seen in the deviation from the straight lines in figure 5.2b). Here, a possible explanation might be the structural changes inside the grease as described by the two effects of Baart et al.'s [5] model, namely shrinking and tilting (cf. section 2.2.1).

So far, both the grease structure and the carrier structure have shown clear influence on oil transition. Beyond this point, additional statements become very vague. Given the fact that the above analysis is based on the paper's nature as a porous medium rather than a structured surface, further research is necessarily required to gain a better understanding of the physics involved. Yet as the transition process happens faster the better the structure can generate capillary forces it can at least be expected that a favorable surface design for oil transport (as discussed in section 3.1) is not unfavorable for the oil transition out of a grease.

5.2. Lubricant transition into tribocontact

5.2.1. Capillary main flow breakdown at the end of the microchannel

So far the capillary flow in a microchannel has been analyzed in situations where the front meniscus has not yet reached the end of the channel. In the absence of the discussed effects which may lead to a complete stop of lubricant advancement or even to flow reversal (e.g., the Marangoni effect or the motion of the microchannel itself), the lubricant continues to flow forward infinitely: Under the assumptions made in chapter four, the capillary front meniscus is expected to keep its shape so that the capillary force is always constant. The increasing length of the fluid column adds viscous resistance and the capillary flow consequently slows down gradually but never stops.

Once the front meniscus reaches the channel end, it breaks down: The air in front of the meniscus gets pushed out of the channel by the lubricant, so that a pressure-drop generating liquid-air interface does only exist at the upper channel surface. As without front meniscus the pressure gradient inside the fluid column vanishes (external effects excluded) and pressure becomes ambient pressure everywhere, the upper meniscus cannot maintain its curved shape neither. Instead, it becomes flat as a consequence of both Young-Laplace equation (2.63) and Gibbs criterion (2.17), cf. figure 5.3b). In summary, the capillary flow has come to a complete stop.

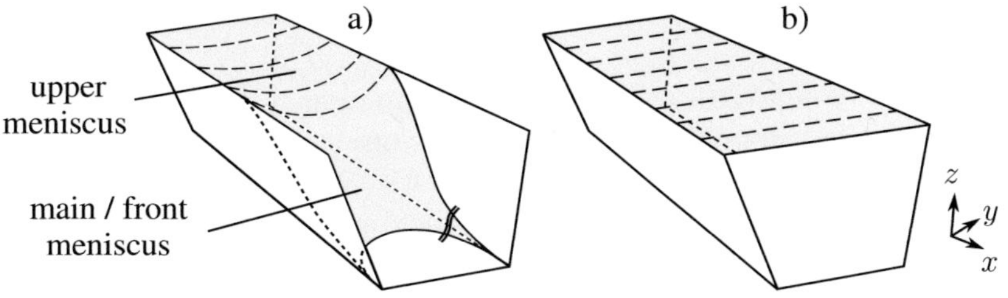

Figure 5.3.: Channel end effect: a) Capillary flow having not yet reached channel end (same as figure 4.7); b) breakdown of capillary flow with vanishing menisci at channel end

Microchannel end within tribocontact

In order to avoid such a flow stop before the lubricant has reached the tribocontact, one might be tempted to simply let the microchannels end within the tribocontact. This yet comes with complications:

First, structures inside the tribocontact might interfere with the function of the tribocontact itself, especially with respect to the pressure buildup in the lubricant film separating the contact partners. An introductory overview of surface design within the tribocontact is given e.g., by Etsion [66] with a fundamental idea visualized in figure 1.2. Based on this a less problematic or even beneficial design should be achievable for example by structuring the channels perpendicular to the contact partners' relative motion in order to generate an additional pressure contribution.

What is more critical is the fact that once the lubricant fluid column has connected with the lubricant fluid inside the contact, the front meniscus and thereby the capillary force equally disappear – in analogy to what happens at the microchannel's end as described above. In addition to that, the pressure in the contact is higher than the ambient pressure typically by several orders of magnitude, which should result in a strong Poiseuille flow away from the tribocontact. This is true if other possibly contributing flow components as summarized for example in Reynolds equation (2.38) can be neglected. In fact, a counterexample to such a simplification is the classical (elasto-) hydrodynamic lubrication theory itself where flow through the tribocontact faces a strong positive pressure gradient before reaching the maximum pressure as seen for example in figure 5.6. Yet from above, normal tribocontact operation may be significantly disturbed by microchannels. As a consequence, a more detailed analysis of the contact including supply structures seems inevitable with some more thoughts given in section 5.2.3.

Microchannel end at tribocontact's boundary

Alternatively, Poiseuille counterflow inside the microchannel may generally be avoided if the channel ends before the tribocontact's pressure buildup starts i.e., at the tribocontact's pressure boundary. Then, the pressure field inside the microchannel (which develops after the capillary flow has filled the channel completely) is uniform at ambient pressure with the capillary filled as seen in figure 5.3b). Again, section 5.2.3 will provide some more details.

5.2.2. Mechanisms of lubricant removal out of the microchannel

So far the result is simple: Even though microchannels have proven capable of transporting lubricants by a capillary force over a large distance (in basic theory even an unlimited distance) they cannot serve the purpose of continuously supplying the tribocontact without an additional mechanism which removes the fluid at the channel end. This is analogously true for any capillary surface structure. If indeed such a removal mechanism did take a certain amount of lubricant out of the channel, then again a capillary front meniscus would form consequently taking care of refilling the capillary channel. The physical nature of the mechanism itself is not important. An extensive summary of potentially exploitable effects is given for example in [178] some of which are discussed in what follows:

Surface tension and inertial force driven flow

Both surface tension driven and inertial force driven flows have been analyzed in their potential to hinder or stop a capillary flow in sections 4.2 - 4.5. Reversing the acting direction of both forces in theory allows to drive a flow forward after the breakdown of a capillary front meniscus. Yet following the argumentation in section 4.4.1 that a surfactant-based Marangoni effect is not likely to be significant (and even less likely to be exploitable) in tribological environment and the thermal Marangoni effect always draws lubricants away from the tribocontact because temperature is higher there than in the surroundings, surface tension driven flow is not seen as a probable solution, even though its potential to remove a liquid out of a reservoir has been demonstrated [68, 145]. This is different for an inertial force driven flow: As long as the movement of the structured body – either coincidentally or intentionally – is responsible for a centrifugal, Euler, and/or translational acceleration force component acting in direction towards the tribocontact, the lubricant may well leave the capillary channel. Figure 5.4 shows oil sucked out of previously filled microchannels after centrifugation. Even though the microchannels end halfway along the test piece, the fluid flow in the continuing slender fluid columns should allow to treat it mathematically in a similar way as if the microchannels were continued, yet without front meniscus pressure drop.

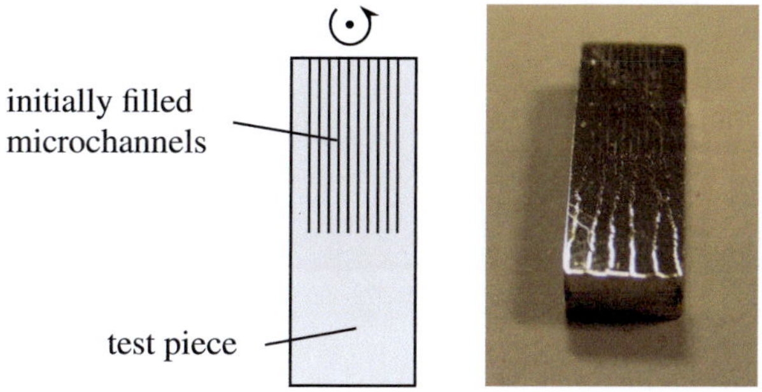

Figure 5.4.: Structured microchannels end at half length of the test piece; channels are initially filled with oil as seen in figure 5.3b); subsequent centrifugation as described in [29] leads to oil flow out of the channels (right image)

Also solid body vibrations transmitted to the fluid column may be a viable alternative. As argued in section 4.5 their computational treatment differs considerably from the modeling framework developed so far – mainly due to inertial contributions inside the fluid column. In practice they yet may combine well with a capillary microchannel transport: Even if they are not directed and only transfer oil out of the microchannel and into the tribocontact in a semi-predictable or statistical way, the removed oil mass immediately leads to a new capillary meniscus in the channel which then draws lubricant to refill the channel.

Flow in capillary substructures

Capillary forces themselves may be present also in the absence of a driving microchannel front meniscus: Especially surface roughness can cause the buildup of driving menisci in channel-like substructures or between post-like structures as seen in figures 3.1a) and 5.5c). Whereas the small size of such menisci leads to remarkable pressure drops from Young-Laplace equation (2.63), force direction and especially high viscous resistance in small gaps must be seen critical. Alternatively, the contacting bodies in the tribocontact together may form edge-like geometries at their contact line which can lead to a corner flow mechanism, cf. section 2.5.3. In any case the same criticism applies as for the microchannel itself: Once the smaller structures have filled there is no reason that transport should continue if the lubricant is not taken away at their front end by another mechanism, such as the following:

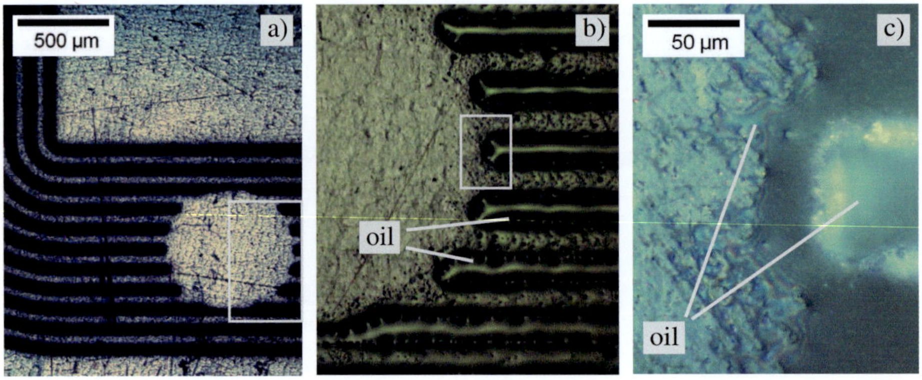

Figure 5.5.: Physics at the end of a microchannel (shown on the same type of test piece with from a) to c) increasing magnification, the white rectangles indicate the location shown in the next magnification step): a) overview of unfilled microchannels, with and without channel end; b) after filling the channels with oil, flow stops at the channel ends; c) surface roughness induced capillary forces pull small quantities of oil out of an ending microchannel

Tribocontact induced Couette flow

An inherent property of any tribocontact is relative motion between the two contacting solid bodies. In fact, even though (elasto-) hydrodynamic lubrication analysis often concentrates on pressure generation, the sole energy source for all effects observed is relative solid body motion (leaving externally pressurized applications aside). The resulting Couette flow transports oil through the tribocontact. Reynolds equation allows to solve for the velocity field inside the contact and on this basis allows to quantify the volumetric flow rate the tribocontact "pumps" from its inlet to its outlet.

If one is able to connect the fluid domain inside a microchannel with the fluid inside the contact e.g., as depicted in figure 5.6a) it must be inferred that tribocontact pumping can be exploited to remove lubricant fluid from the channel. If the mechanism works as desired the tribocontact may even drive the flow in the whole microchannel without the need for a capillary front meniscus. In addition to that, the capability of the effect might even overcome potentially present counteracting menisci if the findings of Powell et al. [158] concerning the removal of a coating fluid out of surface cavities can be transferred to tribological application. For the potential and universal applicability of this mechanism it is discussed in further detail in the following section.

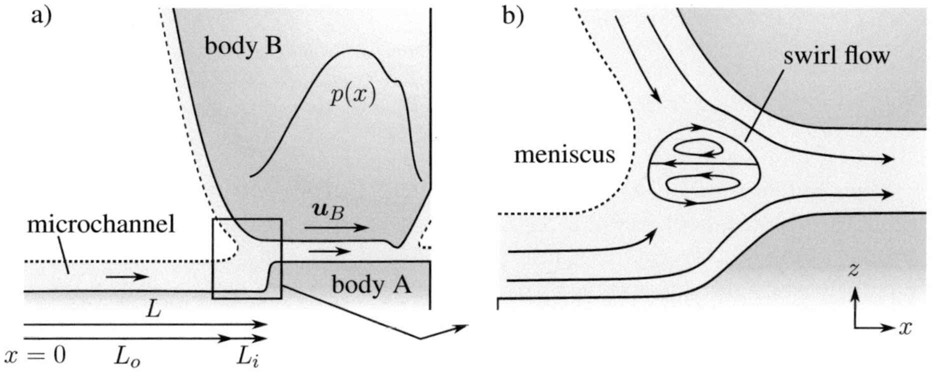

Figure 5.6.: a) EHL contact with characteristic pressure function, cf. [136, 196], adapted to include a supply microchannel; b) tribocontact inlet zone with capillary meniscus and developing swirl flow, adapted from [143]

5.2.3. Tribocontact operation as a driver of microchannel flow

The possibility that a tribocontact generated Couette flow might not only complement but replace a capillarity driven Poiseuille flow in the microchannel is an ambiguous result: On the one hand, the force contribution of a front meniscus is lost for transport, on the other hand, the tribocontact might be able to generate an even higher channel throughput than the capillary force can.

In order to get a first estimate of the relative transporting capability of both mechanisms, one can construct a simple two-dimensional (x, z) model of a microchannel of total length L which is split up into a length L_o outside the tribocontact and a length L_i inside as seen in figure 5.6a). More detailed information about the specific modeling can be found in the appendix. The model can only build in part on the one developed in the previous chapter because the very basic assumption of unidirectional flow cannot be held up any more as is seen in figures 5.6b) and 5.7. The latter visualizes the flow field in transition to the tribocontact under the assumption that the upper surface velocity $u_h(x)$ is externally given. It drives the flow instead of a capillary meniscus. Two examples a) and b) for $u_h(x)$ are shown.

Neglecting the converging tribocontact geometry and concentrating on a simplified rectangular fluid domain instead, it is reasonable to assume that within the contact a no slip boundary condition requires $u_h(x > L_o) \approx \boldsymbol{u}_B \hat{\boldsymbol{e}}_x$, cf. figure 5.6a). Outside, the upper surface velocity is expected to be significantly smaller,

$$u_h\,(x < L_o) \ll u_B \tag{5.1}$$

This leads to lower fluid layers being drawn up while entering the tribocontact. Additionally, a swirl flow is observed in figure 5.7b) in the tribocontact region. Here, fluid parcels in lower fluid layers move in negative x-direction. Obviously, such modeling does not consider the appropriate stress boundary condition at the free liquid surface inside the channel and can only be considered a rough estimate under condition (5.1). Nevertheless, it shows that the lubricant throughput is higher when the microchannel stretch outside the tribocontact is relatively short compared to the length within the contact.

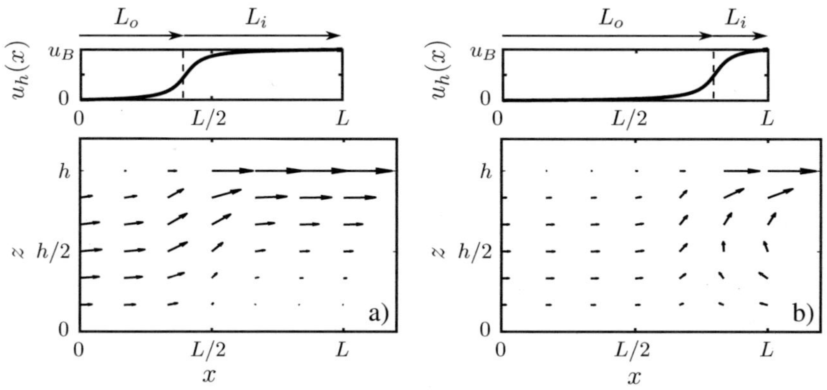

Figure 5.7.: Flow field $\boldsymbol{u}(x, z)$ in the microchannel depending on the fractions of channel length inside (L_i) and outside (L_o) the tribocontact with upper surface velocity $u_h(x)$, which drives the flow, given above; no capillary contribution; computation based on the model given in the appendix; absolute flow rate q is higher in a)

This is confirmed by fully three-dimensional Navier-Stokes flow computations with the help of a commercial finite element software code (COMSOL) as seen in figure 5.8. A basic rectangular channel is split up into a first region with a free surface and a second region where a constant upper surface velocity $u_h = u_B$ is set. In order to compare the results with the two-dimensional model above, the no slip boundary conditions at the vertical channel walls in a) have been replaced in b) - e) by no stress conditions. Clearly, the overall fluid velocity in the channel decreases when reducing the upper surface area where the velocity $u_h = u_B$ is given. The upward-drawn liquid in the tribocontact region can best be seen in e) which is the streamline plot of d). Subfigure f) models a cylinder (body B) rotating with ω_y as part of the tribocontact which draws lubricant out of a channel segment. In addition to the above, a Marangoni effect pulls the liquid away from the contact. If one accepts the model's considerable simplification that shape and location of the free surface meniscus are not an outcome of the model but have been defined, the results fit to what was seen previously: As in figure 4.14 with capillary flow, the tribocontact's pumping ability depends on lower fluid layers to supply the tribocontact whereas upper fluid layers are drawn away from the contact.

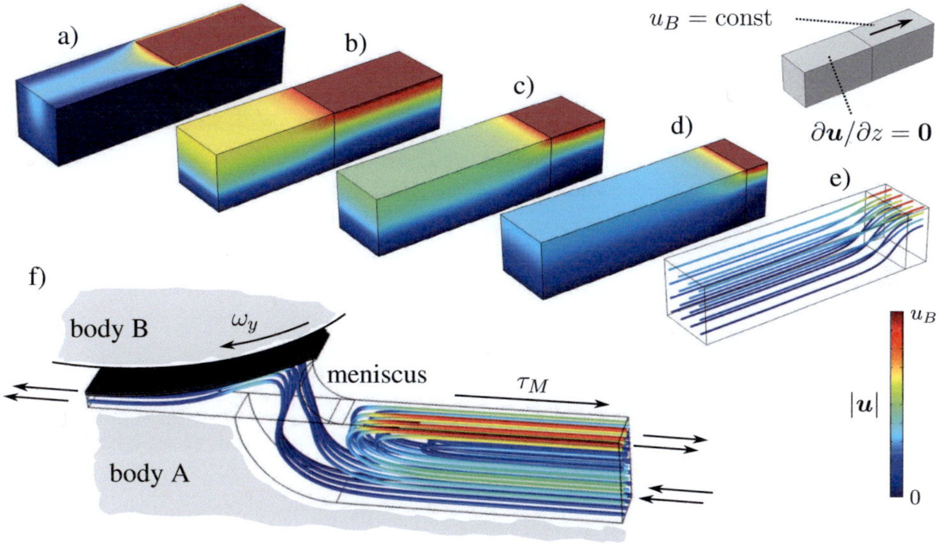

Figure 5.8.: Finite element computation of a Couette flow in a microchannel driven by the movement of a fraction of the channel's total upper surface: a) - e) pure channel flow, f) model of a tribocontact geometry including an influence of the Marangoni effect; see text for more details

To compare the supply capabilities of capillary and tribocontact driven flows, the two-dimensional model is again taken as the basis. The associated flow rate in the two-dimensional microchannel (unit: m^2/s) can be computed as

$$q = \frac{h}{2L} \int_0^L u_h(x) \mathrm{d}x \approx \frac{u_B h}{2} \frac{L_i}{L} \tag{5.2}$$

with details stated in the appendix. The approximation of the integral is based on the assumption that while entering the tribocontact $u_h(x)$ rises rather sharply from a small value outside the contact to u_B inside. The location where this happens is marked by the dashed line in figure 5.7.

Eq. (5.2) shows the proportionality of downstream flow rate q and the fraction of the channel inside the contact (L_i/L). As microchannel design can influence channel length L but not tribocontact dimensions (i.e., design possibilities with respect to L_i are limited), the more decisive result is the inverse proportional relationship of flow rate and channel length $q \propto 1/L$. In this, the microchannel flow driven by a tribocontact is in perfect agreement with the standard Washburn case of capillary channel flow, cf. table 4.7, case A.

Based on the assumptions made, this statement can only be considered an estimate which needs further model refinement and a rigorous augmentation to three dimensions. Especially the channel design at the contact has to be considered in more detail: While eq. (5.2) argues for maximizing the capillaries' length inside the tribocontact, the reasoning in section 5.2.1 (regarding pure removal mechanisms) was opposite due to the sharply increasing pressure towards the center of the contact. For both cases it is necessary to determine the location of the boundary of the tribocontact, at least in its inlet region.

Location of tribocontact's inlet boundary

Especially in the relevant case of starved lubrication the tribocontact boundary's exact geometrical location is still to a considerable degree unclear, both from experimental and modeling perspectives:

EHL theory comes with an array of boundary conditions, for example Swift-Stieber, Prandtl-Hopkins, Coyne-Elrod, or Jakobsson-Floberg-Olsson. The main motivation for the specific choice is generally a better explanation of the physics inside the tribocontact. The questions what exactly happens at the boundary and where the boundary is located are of secondary importance and are answered only vaguely. For measurement difficulties of the exact position of the inlet meniscus, researchers have even switched from location based to volume fraction based inlet modeling [39] when computing Reynolds equation.

Experimental studies have been conducted predominantly with the ball-on-spinning-disk arrangement where a rotating transparent disk allows to analyze the lubricant flow at the contact with a running ball, see for example [131, 195]. Interferometry is used to determine the film height while pressure is measured via micro-transducers [143]. In the static case without disk rotation, capillary forces are responsible for the formation of a circular liquid collar around the tribocontact as seen in figure 5.9. The lubricant-filled area is significantly larger than Hertzian theory's contact circle. Given characteristic fluid domain dimensions are smaller than the capillary length, eq. (2.1), the collar's ground diameter d_L is a function of the material specific contact angles $\theta_{eq,A}$ and $\theta_{eq,B}$

and the liquid volume only and thus can be determined geometrically with relative ease.

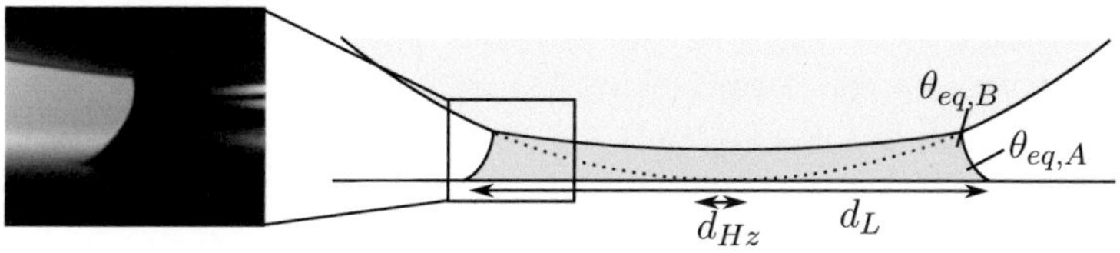

Figure 5.9.: Fluid-filled domain around the static ball - plate contact

Dynamically, fluid flow into and out of the contact area and especially the interplay of capillary forces around the tribocontact and the various fluid dynamical effects make the circular area converge – both in pressure and interferometry measurement – to an area described as approximately "horseshoe" [143] or "butterfly"-shaped [136]. The variation in the basic shape reported is considerable and substantially depends on operating conditions as seen in figure 5.10. If in this case the current knowledge is still limited, even less research results are available for other contact geometries than ball–plate; with considerable uncertainty one might expect a similar behavior.

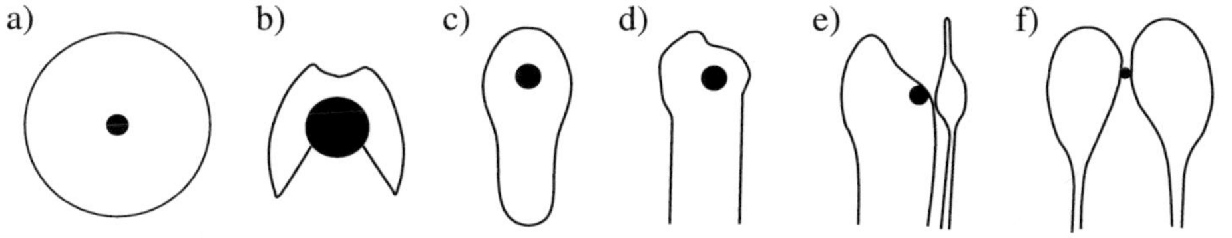

Figure 5.10.: Schematic representation of lubricant-filled area around the Hertzian ball–plate contact (in black) at different operating conditions: a) static case, b)-f) dynamic case based on interferometry and pressure measurements under different operating conditions in b) [143], c) [195] d) and e) [131] f) [160]

Thus from a design perspective robust functioning requires the microchannels to extend safely into the tribocontact's fluid domain. As a consequence, tribocontact fluid dynamics have to be taken into consideration when designing lubricant supply structures:

Towards a comprehensive model

A comprehensive model requires the coupled consideration of (elasto-) hydrodynamic lubrication and microchannel flow theories in three dimensions and over two to three scales: It is seen as a vital area for further research.

From above, the operating tribocontact cannot be treated as a simple pressure or flow rate boundary condition for microchannel flow: The former would predict reverse channel flow away from the contact as the pressure there is much higher than the ambient pressure. With respect to the flow rate, it is not only determined by the tribocontact alone; as has been seen, the overall microchannel length is decisive as additional length adds viscous friction and consequently slows down the channel flow. In the extreme case, any sufficiently long capillary can virtually stop every tribocontact's pumping capabilities.

On the other hand, extending Reynolds equation from two scales (i.e., film height and lateral tribocontact dimensions) to a third one (microchannel length) and to additional boundary conditions (e.g., location and nature of free surface boundaries) should prove at least computationally expensive if not troublesome.

Existing research has developed models for a similar context in two dimensions (x, z), which are – unlike Reynolds equation – not any more based solely on lubrication theory, eq. (2.37), but at least in part on Stokes equations (2.26) or even the full Navier-Stokes momentum equation set (2.20). Instead of focusing on the tribocontact inlet, Bruyère et al. [28] model the split up of the lubricant film in the outlet region of a tribocontact with the help of a Navier-Stokes based phase field ansatz which as true two-phase flow model promises to consider capillary effects. With their model, they are able to reproduce swirl flows (see figure 5.6) which have already been proposed earlier in [191] in the inlet region of the contact. While Mohammadpour et al. [143] point out that as a consequence only part of the lubricant inlet flow is entrained into the contact, these flows may also interfere with the removal of lubricant liquid out of a supply channel. To be able to pay special attention to meniscus regions, Taroni et al. [184] decompose the computation domain: While the tribocontact as an outer region is still modeled with the help of lubrication theory, the inner meniscus region requires a Stokes ansatz before coupling both with the help of matched asymptotic expansions. Generally, the models mentioned underline the influence of capillarity as expressed in the capillary number.

Such logic of a two-dimensional model, if successfully applied to the tribocontact's inlet, will need further extension to cover lateral flows in the x-y plane which might be instrumental in taking liquid out of a microchannel in tribocontact's inlet region. This would as well cover all so-called "in-contact" replenishment flows around the tribocontact i.e., from outlet back to inlet [32, 104]. While knowledge about their nature is limited, their importance cannot be underestimated: If one simply computes the flow rate through the contact by the help of Reynolds equation alone and assumes that every fluid parcel transits contact only once, then for a given amount of lubrication liquid initially present (e.g., bound in a grease drop) the theoretical time the tribocontact is lubricated may well be smaller than the real product's lifetime by several orders of magnitude. Lubricant "recycling" significantly reduces the need for the capillary supply of new lubricant. It might itself be supported by microchannels from the tribocontact's outlet to its inlet.

It cannot be judged if surface curvature, as investigated for example in [197], is relevant. The Marangoni effect yet does not only pull liquid away from the contact in a supply channel, it is furthermore expected to influence near-meniscus flow at the tribocontact inlet, cf. eq. (2.59). Further analysis will require knowledge of temperature gradients at the meniscus. As even shear heating at the inlet is reported to be very well relevant for tribocontact lubrication [86, 153], the unsubstantiated neglect of a possible Marangoni influence does not seem advisable. It cannot be answered at this point if this requires the inclusion of an energy equation into a general model.

Generally the importance of capillary effects in the direct vicinity of the tribocontact is clearly recognized [72] even in the isothermal case and research is ongoing. Even in the case they prevent the tribocontact from "pumping", alternatives with respect to lubricant removal from a transport structure exist as described in section 5.2.2, which offers more freedom in design.

5.2.4. Experimental approach to lubricant transition into tribocontact

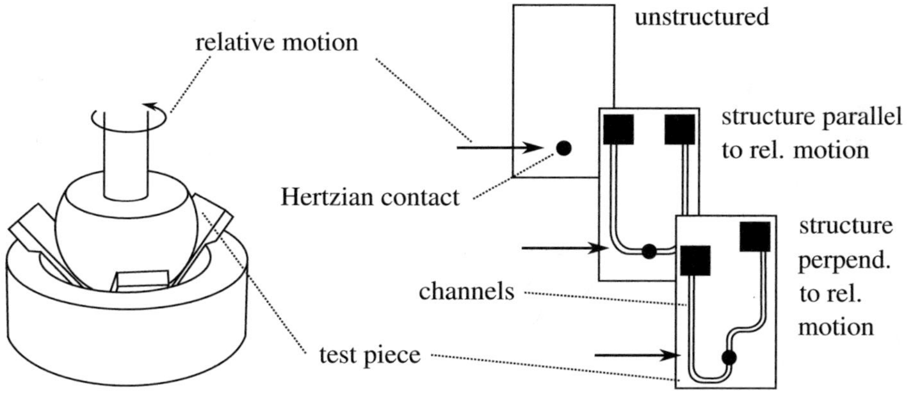

Figure 5.11.: Basic experimental setup to test capillary supply effectiveness of a tribocontact; experiments were conducted with three different sets of test pieces with $b_c = 50$ µm, $\Lambda \approx 1$ and further dimensions similar to figure 3.7

While the previous sections have shown that the physics of lubricant transition out of the capillary transport structure into the tribocontact is still subject to several open questions, the goal of the preliminary experimental approach presented in the following is to demonstrate the general effectiveness of the system as a whole, i.e., an operating tribocontact, which enhances its state of lubrication by serving itself from supply capillaries.

Basis for the experiments is a tribometer in a "ball on three plates" configuration (MCR301 with extension T-PTD200, Anton Paar, Austria) as depicted in figure 5.11: A ball at the end of a rotating shaft touches three test pieces with a normal contact force of 10.6 N. The contact region on the test pieces is subjected to friction by the passing surface of the rotating ball. In order to include the Marangoni effect, ABS is chosen as ball material which generates high frictional heat on the POM test pieces.

The speed of rotation is ramped up linearly from zero to a maximum of 3000 revolutions per minute and then held constant. Without an additional heat source, friction leads to a temperature increase until at the beginning of the melting temperature range of homopolymer POM (slightly above 170 °C) the test pieces deform: As the ball digs itself into the plates the coefficient of friction rises, more heat is generated so that the tribological system fails rather abruptly. The sharp rise in the friction force is measured by a jump in the torque needed to keep the desired rotational speed as seen in figure 5.12.

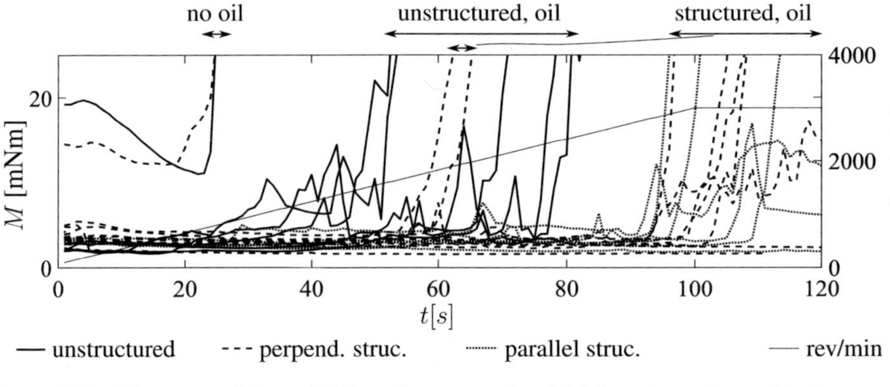

Figure 5.12.: Tribocontact failure of differently structured and lubricated test pieces; friction power is increased over time via rotational speed which causes a temperature rise; failure, if happening, is determined by a jump in moment

The experiment is conducted in various configurations: Unstructured test pieces are compared with laser-structured test pieces as described in section 3.3 yet with adapted geometry so that the channel direction is either parallel or perpendicular to the relative motion in the tribocontact as shown in figure 5.11. Baseline tests without lubricant show early failure irrespective of the use of structured or unstructured plates. Lubrication is done via the same PAO ISO VG 15 as used in previous experiments. As an alternative to capillary filling out of a reservoir, some experiments are conducted with test pieces and/or ball slightly wetted by means of an oil soaked piece of cloth. The individual method of lubrication does not lead to significantly different results. In contrast, structured plates in general fail significantly later than unstructured ones (with the exception of two experiments of plates with perpendicular structure which were roughly able to support a load comparable to the unstructured plates) demonstrating the microchannels' effectiveness.

A point of criticism is the high variance – even among unstructured plates – which needs to be addressed in more detail. Furthermore, as the microchannels cross tribocontact instead of ending at its boundary, the experiment cannot separate between the microchannels' effect as oil supply channels versus the structure's ability to favorably change pressure generation inside the contact itself. Only the first effect is of interest here. If the second effect was dominant, one would expect different results depending on the structure's direction: The multiple converging-diverging geometry features of channels perpendicular to motion direction, cf. figure 1.2, should be clearly superior to channels parallel to motion direction which may well facilitate pressure loss in the contact. Yet the results do not show such a difference – one might even consider the parallel channels' results to be slightly better.

In summary, the experiments suggest that indeed the microchannel oil supply is the reason for the superior tribocontact performance. A more rigorous argument needs further experimental design refinement (e.g., with test pieces not structured through the contact as seen in figure 5.5) which is left open for further research.

6. Conclusion

Structuring surfaces in order to generate capillary flows very well serves the purpose of transporting lubrication liquids from a distant reservoir e.g., a grease towards tribocontact. Based on the greater motivation to counter starvation effects by better supplying tribocontacts with lubricants from outside, concentration neither lay on "in contact" lubricant flows nor on "in contact" surface structuring.

A geometrically rather simple array of constant cross-section microchannels was identified as favorable structure type, due to its superiority regarding strength and direction of capillary force buildup on a given surface area. Fabrication and measurement of slender channels with widths as low as 20 µm, intended to least possibly interfere with the structural integrity of the carrier part, proved challenging on typical tribological materials such as metals or engineering plastics which resulted in lower geometrical accuracy compared to what is known from etching processes in silicon for example. Depending on the application case, the best results were achieved by micromilling or femtosecond laser ablation.

Flow modeling strategy built on the above: While the overall geometrical simplicity of a constant cross-section channel allows for significant simplification in the physical model culminating in the assumption of unidirectional flow, cross-sectional shape and surface roughness needed additional attention. The underlying idea of so-called lubrication theory i.e., two different characteristic length scales, was adapted to microchannel flow by a nondimensionalization procedure. In the base case this led to a classical Poisson type equation for microchannel flow.

Inertial forces from rotating or accelerating solid bodies into which the structure is cut could be included rather simply as body force equivalents. The non-isothermal temperature field around a tribocontact, simply because the contact generates heat, mainly has two consequences: First, implications of viscosity variation along the channel on flow behavior and channel throughput had to be covered. Second, the relevance of the Marangoni effect required surface tension gradients at the open channel surface to be considered.

To include the capillary meniscus at this free surface, simplifying assumptions had to be made and were discussed. A corner flow influence was argued to be negligible. Both approaches pursued in literature which still allow to treat the front meniscus as a flow boundary condition, relying on interfacial pressure drop and the minimization of Gibbs free energy, respectively, show complications with respect to nonstandard cross-sectional channel geometries. They were consequently adapted to separate size and material property influences from characteristic shape influence with the requirement to determine the latter in a one-time lab experiment. In order to make it redundant, future research is needed regarding shape and resulting capillary force generation of a three dimensional meniscus including flow dynamics. Potential model starting points were given, which after successful implementation are expected to come with heavily increased computational cost.

Lab experiments with a variety of structured test pieces were designed and conducted in order to validate the channel flow model. Including non-isothermal conditions, a fine agreement could be reached.

Computation and experiment show that both inertial forces due to moving structured bodies as well as thermal effects can be highly relevant in their potential to draw lubricants away from a tribocontact – inside or outside a microchannel. The buildup of capillary force in the channel yet exercises a strong counteraction. It is for example able to maintain a positive net lubricant flow towards tribocontact due to flow near channel bottom even though upper fluid layers are drawn away from tribocontact by Marangoni induced stresses.

While transport in the capillary channel array itself is, based on the above, well computable, it is the transition zones into and out of the capillaries which come with a variety of open questions: Neither a grease's oil bleeding behavior nor lubricant flow inside and in the immediate vicinity of tribocontact is being understood to a degree required to model the physics in these regions. Especially the connection with tribocontact is important, because if no appropriate transition mechanism is available, capillary channel transport can in the worst case break down completely. On the other hand, tribocontact itself may undertake this transition task as both theoretical considerations and experimental results suggest. From a modeling point of view it has to be asked if interfaces between the respective domains can sensibly be defined or if alternatively they have to be treated as a connected system. First thoughts and approaches have been stated, yet leave plenty of space for coming research.

All in all, capillary surface structures have proven as a reliable means to transport liquid lubricants over comparatively long distances in spite of potentially strong counteracting forces. The channel transport itself is predictable. Lubricant supply of a tribocontact with the help of such structures can be considered a promising new option to enhance tribocontact operation.

Appendix

A. The material derivative in inertial and noninertial reference frames

The absolute velocity of a fluid parcel in an inertial frame of reference (I) is given by ${}_I\boldsymbol{u}$. The relative velocity of the same fluid parcel (i.e., the velocity relative to a non-inertial frame (O)) is ${}_O\boldsymbol{u}$. Both can be expressed as material derivatives of the respective position vectors (cf. figure 2.6):

$$
{}_I\boldsymbol{u} = \frac{{}^I\mathrm{D}\breve{\boldsymbol{s}}}{\mathrm{D}t} \tag{A.1}
$$

$$
{}_O\boldsymbol{u} = \frac{{}^O\mathrm{D}\boldsymbol{x}}{\mathrm{D}t} \tag{A.2}
$$

The main coordinate system of frame (O) is chosen to be fixed at the microchannel's inlet. The channel itself is fabricated into a moving rigid body. Although not necessary it may be convenient to define an auxiliary coordinate system with origin M which is also located on or in the body and moves with (O). For this reason $\breve{p}$ is constant in (O) so that ${}^O\mathrm{D}\breve{\boldsymbol{p}}/\mathrm{D}t = \mathbf{0}$. With $\boldsymbol{x} = -\breve{\boldsymbol{p}} + \breve{\boldsymbol{q}}$ eq. (A.2) yields:

$$
{}_O\boldsymbol{u} = \frac{{}^O\mathrm{D}\breve{\boldsymbol{q}}}{\mathrm{D}t} \tag{A.3}
$$

The velocity of M measured in the inertial frame is:

$$
\frac{{}^I\mathrm{d}\breve{\boldsymbol{r}}}{\mathrm{d}t} = \frac{{}^I\mathrm{D}\breve{\boldsymbol{r}}}{\mathrm{D}t} \tag{A.4}
$$

The fluid parcel velocity expressed in the inertial frame is the sum of the velocity of M, the velocity component due to rotation of the rigid body, and the fluid parcel velocity measured in the moving frame:

$$ {}_I\boldsymbol{u} = \frac{{}^I\mathrm{D}\breve{\boldsymbol{s}}}{\mathrm{D}t} = \frac{{}^I\mathrm{D}\breve{\boldsymbol{r}}}{\mathrm{D}t} + \boldsymbol{\omega} \times \breve{\boldsymbol{q}} + \frac{{}^O\mathrm{D}\breve{\boldsymbol{q}}}{\mathrm{D}t} \tag{A.5} $$

With $\breve{\boldsymbol{s}} = \breve{\boldsymbol{r}} + \breve{\boldsymbol{q}}$:

$$ \frac{{}^I\mathrm{D}\breve{\boldsymbol{q}}}{\mathrm{D}t} = \frac{{}^O\mathrm{D}\breve{\boldsymbol{q}}}{\mathrm{D}t} + \boldsymbol{\omega} \times \breve{\boldsymbol{q}} \tag{A.6} $$

The procedure which resulted in this equation, valid specifically for vector $\breve{\boldsymbol{q}}$, can be generalized to apply for an arbitrary vector $\boldsymbol{o}$ to give the basic equation:

$$ \frac{{}^I\mathrm{D}\boldsymbol{o}}{\mathrm{D}t} = \frac{{}^O\mathrm{D}\boldsymbol{o}}{\mathrm{D}t} + \boldsymbol{\omega} \times \boldsymbol{o} \tag{A.7} $$

In the special case that a vector is aligned with $\boldsymbol{\omega}$ i.e., $\boldsymbol{\omega}$ itself, the cross product vanishes and consequently the material derivative of such a vector is invariant to the reference frame chosen. With respect to $\boldsymbol{\omega}$, the material and time derivatives are equal, so that:

$$ \frac{{}^I\mathrm{D}\boldsymbol{\omega}}{\mathrm{D}t} = \frac{{}^O\mathrm{D}\boldsymbol{\omega}}{\mathrm{D}t} = \frac{\mathrm{D}\boldsymbol{\omega}}{\mathrm{D}t} = \frac{\mathrm{d}\boldsymbol{\omega}}{\mathrm{d}t} \tag{A.8} $$

Now applying the material derivative in an inertial frame ${}^I\mathrm{D}/\mathrm{D}t$ to eq. (A.5):

$$ \frac{{}^I\mathrm{D}_I\boldsymbol{u}}{\mathrm{D}t} = \frac{{}^I\mathrm{D}^2\breve{\boldsymbol{s}}}{\mathrm{D}t^2} = \frac{{}^I\mathrm{D}^2\breve{\boldsymbol{r}}}{\mathrm{D}t^2} + \frac{{}^I\mathrm{D}}{\mathrm{D}t}(\boldsymbol{\omega} \times \breve{\boldsymbol{q}}) + \frac{{}^I\mathrm{D}}{\mathrm{D}t}\left(\frac{{}^O D\breve{\boldsymbol{q}}}{Dt}\right) \tag{A.9} $$

With eq. (A.7) and applying the product rule yields:

$$ \frac{{}^I\mathrm{D}_I\boldsymbol{u}}{\mathrm{D}t} = \frac{{}^I\mathrm{D}^2\breve{\boldsymbol{r}}}{\mathrm{D}t^2} + \frac{{}^I\mathrm{D}\boldsymbol{\omega}}{Dt} \times \breve{\boldsymbol{q}} + \boldsymbol{\omega} \times \frac{{}^I D\breve{\boldsymbol{q}}}{\mathrm{D}t} + \frac{{}^O\mathrm{D}^2\breve{\boldsymbol{q}}}{\mathrm{D}t^2} + \boldsymbol{\omega} \times \frac{{}^O D\breve{\boldsymbol{q}}}{\mathrm{D}t} \tag{A.10} $$

where

$$ \boldsymbol{\omega} \times \frac{{}^I\mathrm{D}\breve{\boldsymbol{q}}}{\mathrm{D}t} = \boldsymbol{\omega} \times \frac{{}^O\mathrm{D}\breve{\boldsymbol{q}}}{\mathrm{D}t} + \boldsymbol{\omega} \times (\boldsymbol{\omega} \times \breve{\boldsymbol{q}}) \tag{A.11} $$

from eq. (A.7).

Combining eq. (A.8), (A.10), and (A.11) leads to:

$$\frac{{}^{I}\mathrm{D}_{I}\boldsymbol{u}}{\mathrm{D}t} = \frac{{}^{I}\mathrm{D}^{2}\breve{\boldsymbol{r}}}{\mathrm{D}t^{2}} + \frac{\mathrm{d}\boldsymbol{\omega}}{\mathrm{d}t} \times \breve{\boldsymbol{q}} + 2\boldsymbol{\omega} \times \frac{{}^{O}\mathrm{D}\breve{\boldsymbol{q}}}{\mathrm{D}t} + \boldsymbol{\omega} \times (\boldsymbol{\omega} \times \breve{\boldsymbol{q}}) + \frac{{}^{O}\mathrm{D}^{2}\breve{\boldsymbol{q}}}{\mathrm{D}t^{2}} \tag{A.12}$$

With eq. (A.3) and (A.4):

$$\frac{{}^{I}\mathrm{D}_{I}\boldsymbol{u}}{\mathrm{D}t} = \frac{{}^{O}\mathrm{D}_{O}\boldsymbol{u}}{\mathrm{D}t} + \boldsymbol{\omega} \times (\boldsymbol{\omega} \times \breve{\boldsymbol{q}}) + 2\boldsymbol{\omega} \times {}_{O}\boldsymbol{u} + \frac{\mathrm{d}\boldsymbol{\omega}}{\mathrm{d}t} \times \breve{\boldsymbol{q}} + \frac{{}^{I}\mathrm{d}^{2}\breve{\boldsymbol{r}}}{\mathrm{d}t^{2}} \tag{A.13}$$

B. Simple two-dimensional model of channel flow driven by variable upper surface velocity

A channel of unlimited width can be modeled in two dimensions, x and z, because the flow pattern does not change in lateral direction. Figure 5.7 shows the flow in a channel segment with length L and constant height h driven by a given upper surface velocity $u_h(x)$. Based on lubrication theory, eq. (2.37) is taken as the starting point for the model derivation. In two dimensions it reads:

$$\frac{\partial p}{\partial x} = \eta \frac{\partial^2 u}{\partial z^2} \tag{B.1}$$

$$\frac{\partial p}{\partial z} = 0 \tag{B.2}$$

$$\frac{\partial u}{\partial x} + \frac{\partial w}{\partial z} = 0 \tag{B.3}$$

The viscosity η is assumed to be constant. From eq. (B.2) the pressure does not vary in z-direction, so that $p = p(x)$. With this, integrating eq. (B.1) twice with respect to z yields:

$$u(x,z) = \frac{1}{2\eta}\frac{\partial p}{\partial x} z^2 + c_1 z + c_2 \tag{B.4}$$

Here, c_1 and c_2 are constants of integration which can be determined by applying the appropriate boundary conditions for horizontal flow (no slip at the channel bottom and the given velocity at the upper fluid surface):

$$u(z{=}0) = 0 \tag{B.5}$$
$$u(z{=}h) = u_h(x) \tag{B.6}$$

Then eq. (B.4) becomes:

$$u(x,z) = \frac{1}{2\eta}\frac{\partial p}{\partial x} z\,(z-h) + u_h \frac{z}{h} \tag{B.7}$$

After differentiation with respect to x one can apply the continuity equation (B.3):

$$\frac{\partial w}{\partial z} = -\frac{\partial u}{\partial x} = -\frac{1}{2\eta}\frac{\partial^2 p}{\partial x^2} z\,(z-h) - \frac{\partial u_h}{\partial x}\frac{z}{h} \tag{B.8}$$

Integration with respect to z yields:

$$w(x,z) = -\frac{1}{6\eta}\frac{\partial^2 p}{\partial x^2} z^3 + \frac{1}{4\eta}\frac{\partial^2 p}{\partial x^2} z^2 h - \frac{1}{2}\frac{\partial u_h}{\partial x}\frac{z^2}{h} \tag{B.9}$$

Here, the constant of integration c_3 has already been omitted because the first of the following two boundary conditions for vertical flow directly leads to $c_3 = 0$:

$$w(z{=}0) = 0 \tag{B.10}$$
$$w(z{=}h) = 0 \tag{B.11}$$

Applying eq. (B.11) to eq. (B.9) yields:

$$\frac{\partial^2 p}{\partial x^2} = \frac{6\eta}{h^2}\frac{\partial u_h}{\partial x} \tag{B.12}$$

This result could have been directly derived from Reynolds equation (2.38), with $u_h = u_B$ and $u_A = 0$. Disregarding lateral contributions for the two-dimensional nature of this model and without the squeeze and local expansion terms (which are not relevant due to the assumptions made) one is left with a balance of a Poiseuille and a Couette term, namely eq. (B.12). Double integration with respect to x leads to:

$$p(x) = \frac{6\eta}{h^2}\int u_h(x)\,\mathrm{d}x + c_4 x + c_5 \tag{B.13}$$

As in the rest of this work the pressure level outside the channel is ambient pressure p_{am}. Here, it directly defines the boundary conditions both at the channel inlet and at the channel outlet, because this model purely relies on shear-driven flow and no capillary contribution is included:

$$p(x{=}0) = p_{am} \tag{B.14}$$
$$p(x{=}L) = p_{am} \tag{B.15}$$

With these conditions the constants of integration $c_4 = -(6\eta)/(h^2 L) \int_0^L u_h(x)\,\mathrm{d}x$ and $c_5 = p_{am}$ can be determined inside the channel's domain (especially for $0 \leq x \leq L$). Eq. (B.13) then becomes

$$p(x) = \frac{6\eta}{h^2}\left(\int_0^x u_h(\zeta)\,\mathrm{d}\zeta - \frac{x}{L}\int_0^L u_h(x)\,\mathrm{d}x\right) + p_{am} \tag{B.16}$$

and can be differentiated with respect to x to yield:

$$\frac{\partial p}{\partial x} = \frac{6\eta}{h^2}\left(u_h(x) - \frac{1}{L}\int_0^L u_h(x)\,\mathrm{d}x\right) \tag{B.17}$$

With eq. (B.7) this allows to compute the downstream velocity field in the channel:

$$u(x,z) = \frac{3}{h^2}\left(u_h(x) - \frac{1}{L}\int_0^L u_h(x)\,\mathrm{d}x\right) z(z-h) + u_h(x)\frac{z}{h} \tag{B.18}$$

Combining eq. (B.9) and (B.12) leads to its vertical counterpart:

$$w(x,z) = \frac{\partial u_h}{\partial x}\frac{z^2}{h^2}(h-z) \tag{B.19}$$

In two dimensions the channel downstream flow rate q, defined in eq. (2.50), reads

$$q(x) = \int_0^h u(x,z)\,\mathrm{d}z \tag{B.20}$$

and is measured in [m²/s]. It can be computed with the help of eq. (B.18):

$$q = \frac{h}{2L} \int_0^L u_h(x)\,\mathrm{d}x \tag{B.21}$$

Based on this equation, the channel downstream flow rate is proportional to the channel height h and the integral of the upper surface velocity u_h and inversely proportional to the channel length L if the integral is held constant.

C. Nondimensionalization

The following list of variables' nondimensionalization contains generally valid formula as well as formula which are different when applied in thin film and microchannel context, respectively:

general		thin film	microchannel
$x^* = x/l_0$	$z^* = z/h_0 = z/(\delta l_0)$	$y^* = y/l_0$	$y^* = y/h_0 = y/(\delta l_0)$
$u^* = u/u_0$	$w^* = w/(\delta u_0)$	$v^* = v/u_0$	$v^* = v/(\delta u_0)$
$t^* = u_0 t/l_0$	$p^* = (h_0^2 p)/(\eta_0 u_0 l_0)$		
$f^* = (h_0^2 f)/(\eta_0 u_0)$	$\boldsymbol{a}_T^* = \boldsymbol{a}_T/a_{0T}$		
$\boldsymbol{\omega}^* = \boldsymbol{\omega}/\omega_0$	$\boldsymbol{\alpha}^* = \boldsymbol{\alpha}/\alpha_0$		
$\tau^* = (h_0 \tau)/(\eta_0 u_0)$	$\eta^* = \eta/\eta_0$ (if not a constant)		
$\nabla^* = l_0 \nabla$	$T^* = T/T_0$		

List of Figures

List of Tables

Bibliography

[1] M. M. Aeinehvand, F. Ibrahim, S. W. Harun, W. Al-Faqheri, T. H. G. Thio, A. Kazemzadeh, and M. Madou. Latex micro-balloon pumping in centrifugal microfluidic platforms. *Lab on a Chip*, 14(5):988–997, 2014.

[2] A. Alexeev, T. Gambaryan-Roisman, and P. Stephan. Marangoni convection and heat transfer in thin liquid films on heated walls with topography: Experiments and numerical study. *Physics of Fluids*, 17:062106, 2005.

[3] F. Ali, I. Křupka, and M. Hartl. Enhancing the parameters of starved EHL point conjunctions by artificially induced replenishment. *Tribology International*, 66:134–142, 2013.

[4] T. Andrukh, D. Monaenkova, B. Rubin, W.-K. Lee, and K. G. Kornev. Meniscus formation in a capillary and the role of contact line friction. *Soft matter*, 10(4):609–615, 2014.

[5] P. Baart, B. van der Vorst, P. M. Lugt, and R. A. van Ostayen. Oil-bleeding model for lubricating grease based on viscous flow through a porous microstructure. *Tribology Transactions*, 53(3):340–348, 2010.

[6] K. Bach. *Lineare und nichtlineare Analyse strömungsinduzierter Schwingungen in schlanken elastischen Kanälen*. PhD thesis, Karlsruher Institut für Technologie (KIT), 2014.

[7] K. Bach, H. Hetzler, and W. Seemann. Oscillations of a channel with a flexible wall conveying viscous flow. *PAMM*, 12(1):245–246, 2012.

[8] L. Baldas and R. Caen. Experimental methods. In S. Colin, editor, *Microfluidics*, chapter 7, pages 303–347. John Wiley & Sons, 2010.

[9] D. Bartel. *Simulation von Tribosystemen*. Vieweg+Teubner, 2010.

[10] O. Bäumchen and K. Jacobs. Can liquids slide? Linking stability and dynamics of thin liquid films to microscopic material properties. *Soft Matter*, 6(24):6028–6035, 2010.

[11] J. Bear. *Dynamics of fluids in porous media*. Dover, 1988 (Corrected reprint).

[12] P. Bergé and M. Dubois. Rayleigh-Bénard convection. *Contemporary Physics*, 25(6):535–582, 1984.

[13] J. D. Bernardin, I. Mudawar, C. B. Walsh, and E. I. Franses. Contact angle temperature dependence for water droplets on practical aluminum surfaces. *International Journal of Heat and Mass Transfer*, 40(5):1017–1033, 1997.

[14] J. Berthier and K. A. Brakke. Wetting of solid walls and spontaneous capillary flow. In K. L. Mittal, editor, *Advances in Contact Angle, Wettability and Adhesion*, volume 2, chapter 1, pages 3–46. John Wiley & Sons, 2015.

[15] J. Berthier, D. Gosselin, and E. Berthier. A generalization of the Lucas–Washburn–Rideal law to composite microchannels of arbitrary cross section. *Microfluidics and Nanofluidics*, 19(3):497–507, 2015.

[16] J. Berthier, D. Gosselin, A. Pham, F. Boizot, G. Delapierre, N. Belgacem, and D. Chaussy. Spontaneous capillary flows in piecewise varying cross section microchannels. *Sensors and Actuators B: Chemical*, 223:868–877, 2016.

[17] J. Bico and D. Quéré. Rise of liquids and bubbles in angular capillary tubes. *Journal of Colloid and Interface Science*, 247(1):162–166, 2002.

[18] J. Bico, U. Thiele, and D. Quéré. Wetting of textured surfaces. *Colloids and Surfaces A: Physicochemical and Engineering Aspects*, 206(1):41–46, 2002.

[19] K. Birdi, D. Vu, A. Winter, and A. Nørregård. Capillary rise of liquids in rectangular tubings. *Colloid and Polymer Science*, 266(5):470–474, 1988.

[20] T. Bjøntegaard and E. M. Rønquist. Simulation of three-dimensional Bénard-Marangoni flows including deformed surfaces. *Communications in Computational Physics*, 5(2-4):273–295, 2009.

[21] O. Bliznyuk, H. P. Jansen, E. S. Kooij, H. J. Zandvliet, and B. Poelsema. Smart design of stripe-patterned gradient surfaces to control droplet motion. *Langmuir*, 27(17):11238–11245, 2011.

[22] C. Bosanquet. LV. On the flow of liquids into capillary tubes. *Philosophical Magazine Series 6*, 45(267):525–531, 1923.

[23] K. A. Brakke. The surface evolver. *Experimental Mathematics*, 1(2):141–165, 1992.

[24] T. Brenner, T. Glatzel, R. Zengerle, and J. Ducrée. Frequency-dependent transversal flow control in centrifugal microfluidics. *Lab on a Chip*, 5(2):146–150, 2005.

[25] D. Brewe and B. Hamrock. Geometry and starvation effects in hydrodynamic lubrication. In *Proceedings of the 59th Symposium of the AGARD Propulsion and Energetics Panel: Problems in Bearings and Lubrication, No. 323*, 1982.

[26] J. P. Brody, P. Yager, R. E. Goldstein, and R. H. Austin. Biotechnology at low Reynolds numbers. *Biophysical Journal*, 71(6):3430–3441, 1996.

[27] H. Bruus. *Theoretical microfluidics*. Oxford Master Series in Condensed Matter Physics. Oxford University Press, 2008.

[28] V. Bruyère, N. Fillot, G. Morales-Espejel, and P. Vergne. A two-phase flow approach for the outlet of lubricated line contacts. *Journal of Tribology*, 134(4):041503, 2012.

[29] F. Bühner. *Untersuchung von Öltransportvorgängen auf laserstrukturierten Oberflächen*. Bachelor's thesis, Hochschule Offenburg, 2012.

[30] T. Cambau, J. Bico, and E. Reyssat. Capillary rise between flexible walls. *EPL (Europhysics Letters)*, 96(2):24001, 2011.

[31] A. Cameron. *The principles of lubrication*. John Wiley & Sons, 1966.

[32] P. Cann, B. Damiens, and A. Lubrecht. The transition between fully flooded and starved regimes in EHL. *Tribology International*, 37(10):859–864, 2004.

[33] F.-M. Chang, S.-J. Hong, Y.-J. Sheng, and H.-K. Tsao. Wetting invasion and retreat across a corner boundary. *The Journal of Physical Chemistry C*, 114(3):1615–1621, 2010.

[34] Y.-J. Chang, S.-C. Chen, and C.-L. Hsu. Study on microchannel design and burst frequency detection for centrifugal microfluidic system. *Advances in Materials Science and Engineering*, 2013, 2013.

[35] J. M. Chen, P.-C. Huang, and M.-G. Lin. Analysis and experiment of capillary valves for microfluidics on a rotating disk. *Microfluidics and Nanofluidics*, 4(5):427–437, 2008.

[36] T. Chen. Capillary force-driven fluid flow of a wetting liquid in open grooves with different sizes. In *Proceedings of the IEEE Intersociety Conference on Thermal and Thermomechanical Phenomena in Electronic Systems (ITherm)*, pages 388–396, 2014.

[37] Y. Chen, L. S. Melvin, S. Rodriguez, D. Bell, and M. M. Weislogel. Capillary driven flow in micro scale surface structures. *Microelectronic Engineering*, 86(4):1317–1320, 2009.

[38] Y.-F. Chen, F.-G. Tseng, S.-Y. ChangChien, M.-H. Chen, R.-J. Yu, and C.-C. Chieng. Surface tension driven flow for open microchannels with different turning angles. *Microfluidics and Nanofluidics*, 5(2):193–203, 2008.

[39] F. Chevalier, P. Cann, F. Colin, G. Dalmaz, and A. Lubrecht. Film thickness in starved EHL point contacts. *Journal of Tribology*, 120(1):126–133, 1998.

[40] S. Chibbaro, E. Costa, D. Dimitrov, F. Diotallevi, A. Milchev, D. Palmieri, G. Pontrelli, and S. Succi. Capillary filling in microchannels with wall corrugations: a comparative study of the concus- finn criterion by continuum, kinetic, and atomistic approaches. *Langmuir*, 25(21):12653–12660, 2009.

[41] S. Colin. *Microfluidics*. John Wiley & Sons, 2010.

[42] P. Comanns, K. Winands, M. Pothen, R. A. Bott, H. Wagner, and W. Baumgartner. The texas horned lizard as model for robust capillary structures for passive directional transport of cooling lubricants. In *Proceedings of the SPIE conference on Smart Structures and Materials + Nondestructive Evaluation and Health Monitoring*, volume 9797, page 979711, 2016.

[43] P. Concus and R. Finn. On capillary free surfaces in the absence of gravity. *Acta Mathematica*, 132(1):177–198, 1974.

[44] E. Contraires, J. Teisseires, M. Guilbert, E. Barthel, and E. Sondergard. Unidirectional wetting of anisotropic textured GaSb surfaces. In *Proceedings of the 40th Leeds-Lyon Symposium on Tribology & Tribochemistry Forum*, 2013.

[45] L. Courbin, E. Denieul, E. Dressaire, M. Roper, A. Ajdari, and H. A. Stone. Imbibition by polygonal spreading on microdecorated surfaces. *Nature materials*, 6(9):661–664, 2007.

[46] G. Croce and P. D'Agaro. Numerical simulation of roughness effect on microchannel heat transfer and pressure drop in laminar flow. *Journal of Physics D: Applied Physics*, 38(10):1518, 2005.

[47] N. Damean and P. Regtien. Poiseuille number for the fully developed laminar flow through hexagonal ducts etched in $\langle\ 1\ 0\ 0\ \rangle$ silicon. *Sensors and Actuators A: Physical*, 90(1):96–101, 2001.

[48] B. Damiens, C. Venner, P. Cann, and A. Lubrecht. Starved lubrication of elliptical EHD contacts. *Journal of Tribology*, 126(1):105–111, 2004.

[49] A. A. Darhuber, J. M. Davis, S. M. Troian, and W. W. Reisner. Thermocapillary actuation of liquid flow on chemically patterned surfaces. *Physics of Fluids*, 15:1295, 2003.

[50] A. A. Darhuber and S. M. Troian. Principles of microfluidic actuation by modulation of surface stresses. *Annual Review of Fluid Mechanics*, 37:425–455, 2005.

[51] A. A. Darhuber, J. P. Valentino, J. M. Davis, S. M. Troian, and S. Wagner. Microfluidic actuation by modulation of surface stresses. *Applied Physics Letters*, 82(4):657–659, 2003.

[52] P.-G. De Gennes. Wetting: statics and dynamics. *Reviews of Modern Physics*, 57(3):827–863, 1985.

[53] P.-G. De Gennes, F. Brochard-Wyart, and D. Quéré. *Capillarity and wetting phenomena: drops, bubbles, pearls, waves*. Springer, 2004.

[54] A. de Lazzer, D. Langbein, M. Dreyer, and H. J. Rath. Mean curvature of liquid surfaces in cylindrical containers of arbitrary cross-section. *Microgravity - Science and Technology*, 9(3):208–219, 1996.

[55] A. De Wit, D. Gallez, and C. Christov. Nonlinear evolution equations for thin liquid films with insoluble surfactants. *Physics of Fluids (1994-present)*, 6(10):3256–3266, 1994.

[56] M. Denny. Tree hydraulics: how sap rises. *European Journal of Physics*, 33(1):43, 2012.

[57] Z. Dimkovski, C. Anderberg, R. Ohlsson, and B.-G. Rosén. Characterisation of cylinder liner honing textures for production control. In R. Leach, editor, *Characterisation of Areal Surface Texture*, chapter 11, pages 281–302. Springer, 2013.

[58] DIN 51817. Testing of lubricants - determination of oil separation from greases under static conditions. Standard, Deutsches Institut für Normung e.V., 2014.

[59] M. Dong and I. Chatzis. The imbibition and flow of a wetting liquid along the corners of a square capillary tube. *Journal of Colloid and Interface Science*, 172(2):278–288, 1995.

[60] M. Dreyer, A. Delgado, and H. J. Rath. Capillary rise of liquid between parallel plates under microgravity. *Journal of Colloid and Interface Science*, 163(1):158–168, 1994.

[61] D. Drikakis and M. Frank. Advances and challenges in computational research of micro- and nanoflows. *Microfluidics and Nanofluidics*, 19(5):1019–1033, 2015.

[62] M. Duarte, A. Lasagni, R. Giovanelli, J. Narciso, E. Louis, and F. Mücklich. Increasing lubricant film lifetime by grooving periodical patterns using laser interference metallurgy. *Advanced Engineering Materials*, 10(6):554–558, 2008.

[63] J. Ducrée, T. Brenner, S. Haeberle, T. Glatzel, and R. Zengerle. Multilamination of flows in planar networks of rotating microchannels. *Microfluidics and Nanofluidics*, 2(1):78–84, 2006.

[64] J. Ducrée, S. Haeberle, T. Brenner, T. Glatzel, and R. Zengerle. Patterning of flow and mixing in rotating radial microchannels. *Microfluidics and Nanofluidics*, 2(2):97–105, 2006.

[65] A. G. Emslie, F. T. Bonner, and L. G. Peck. Flow of a viscous liquid on a rotating disk. *Journal of Applied Physics*, 29(5):858–862, 1958.

[66] I. Etsion. Modeling of surface texturing in hydrodynamic lubrication. *Friction*, 1(3):195–209, 2013.

[67] G. Fang and A. Amirfazli. Understanding the edge effect in wetting: a thermodynamic approach. *Langmuir*, 28(25):9421–9430, 2012.

[68] X. Fanton, A. Cazabat, and D. Quéré. Thickness and shape of films driven by a Marangoni flow. *Langmuir*, 12(24):5875–5880, 1996.

[69] R. Farahi, A. Passian, T. Ferrell, and T. Thundat. Microfluidic manipulation via Marangoni forces. *Applied Physics Letters*, 85(18):4237–4239, 2004.

[70] J. Feng and J. P. Rothstein. One-way wicking in open micro-channels controlled by channel topography. *Journal of Colloid and Interface Science*, 2013.

[71] B. Figliuzzi and C. Buie. Rise in optimized capillary channels. *Journal of Fluid Mechanics*, 731:142–161, 2013.

[72] N. Fillot. *Modélisation numérique multi-échelles des contacts lubrifiés*. Habilitation thesis, INSA de Lyon, 2012.

[73] R. Finn. Capillary surface interfaces. *Notices of the AMS*, 46(7):770–781, 1999.

[74] R. Finn. On the equations of capillarity. *Journal of Mathematical Fluid Mechanics*, 3(2):139–151, 2001.

[75] M. Furrer, L. Saraceno, A. Mariani, and G. P. Celata. Capillary pressure influence on open channels pressure drop. *International Journal of Thermal Sciences*, 70:102–113, 2013.

[76] M. Gad-el-Hak. *The MEMS handbook*. CRC press, 2001.

[77] J. L. Garcia-Cordero, L. Basabe-Desmonts, J. Ducrée, and A. J. Ricco. Liquid recirculation in microfluidic channels by the interplay of capillary and centrifugal forces. *Microfluidics and Nanofluidics*, 9(4-5):695–703, 2010.

[78] N. Garnier, R. O. Grigoriev, and M. F. Schatz. Optical manipulation of microscale fluid flow. *Physical Review Letters*, 91(5):054501, 2003.

[79] P. Gaskell, M. Savage, and H. Thompson. Stagnation–saddle points and flow patterns in Stokes flow between contra-rotating cylinders. *Journal of Fluid Mechanics*, 370:221–247, 1998.

[80] M. Geiger, S. Roth, and W. Becker. Influence of laser-produced microstructures on the tribological behaviour of ceramics. *Surface and Coatings Technology*, 100:17–22, 1998.

[81] T. Glatzel, C. Litterst, C. Cupelli, T. Lindemann, C. Moosmann, R. Niekrawietz, W. Streule, R. Zengerle, and P. Koltay. Computational fluid dynamics (CFD) software tools for microfluidic applications – A case study. *Computers & Fluids*, 37(3):218–235, 2008.

[82] A. Glavan, R. Martinez, E. J. Maxwell, A. B. Subramaniam, R. M. Nunes, S. Soha, and G. M. Whitesides. Rapid fabrication of pressure-driven open-channel microfluidic devices in omniphobic R^F paper. *Lab on a Chip*, 13(15):2922–2930, 2013.

[83] G. Goranovic and H. Bruus. Simulations in microfluidics. In O. Geschke, H. Klank, and P. Telleman, editors, *Microsystem engineering of Lab-on-a-Chip devices*, chapter 5, pages 79–116. Wiley-VCH Verlag, 2004.

[84] E. F. Greco and R. O. Grigoriev. Thermocapillary migration of interfacial droplets. *Physics of Fluids*, 21:042105, 2009.

[85] M. D. Greenberg. *Advanced engineering mathematics*. Prentice-Hall, 2nd edition, 1998.

[86] J. Greenwood and J. Kauzlarich. Inlet shear heating in elastohydrodynamic lubrication. *Journal of Lubrication Technology*, 95(4):417–423, 1973.

[87] E. M. Greitzer, C. S. Tan, and M. B. Graf. *Internal flow: Concepts and Applications*. Cambridge University Press, 2007.

[88] M. Groll, M. Schneider, V. Sartre, M. C. Zaghdoudi, and M. Lallemand. Thermal control of electronic equipment by heat pipes. *Revue générale de thermique*, 37(5):323–352, 1998.

[89] S. Guizard, A. Semerok, J. Gaudin, M. Hashida, P. Martin, and F. Quéré. Femtosecond laser ablation of transparent dielectrics: measurement and modelisation of crater profiles. *Applied Surface Science*, 186(1):364–368, 2002.

[90] C. Guo, X. Hu, T. Wang, and D. Tang. Analysis of axial meniscus jump-like transition in rectangular microgrooves. In *Proceedings of the ASME 8th International Conference on Nanochannels, Microchannels, and Minichannels collocated with 3rd Joint US-European Fluids Engineering Summer Meeting*, pages 697–703, 2010.

[91] L. Guo, H. Xu, and L. Gong. Influence of wall roughness models on fluid flow and heat transfer in microchannels. *Applied Thermal Engineering*, 84:399–408, 2015.

[92] B. J. Hamrock, S. R. Schmid, and B. O. Jacobson. *Fundamentals of fluid film lubrication*. Dekker, 2nd edition, 2004.

[93] M. Hans, F. Müller, S. Grandthyll, S. Hüfner, and F. Mücklich. Anisotropic wetting of copper alloys induced by one-step laser micro-patterning. *Applied Surface Science*, 263:416–422, 2012.

[94] T. Härtig. *Stoffübertragung beim Spritzgießen*. PhD thesis, Technische Universität Chemnitz, 2013.

[95] Q. He and N. Kasagi. Phase-Field simulation of small capillary-number two-phase flow in a microtube. *Fluid Dynamics Research*, 40(7):497–509, 2008.

[96] A. Hemmy. The flow of viscous liquids through slightly conical tubes. *Proceedings of the Physical Society of London*, 34(1):22, 1921.

[97] S. Herb. *Modellbildung und Simulation des Fluidtransports im Mikrokanal im Hinblick auf thermische Einflüsse*. Master's thesis, Karlsruher Institut für Technologie, 2015.

[98] U. Hermens, S. Kirner, C. Emonts, P. Comanns, E. Skoulas, A. Mimidis, H. Mescheder, K. Winands, J. Krüger, E. Stratakis, and J. Bonse. Mimicking lizard-like surface structures upon ultrashort laser pulse irradiation of inorganic materials. *Applied Surface Science*, 2017.

[99] H. Herwig. *Strömungsmechanik*. Vieweg+Teubner, 2008.

[100] B. G. Higgins. Film flow on a rotating disk. *Physics of Fluids*, 29(11):3522, 1986.

[101] K. Holmberg, P. Andersson, and A. Erdemir. Global energy consumption due to friction in passenger cars. *Tribology International*, 47:221–234, 2012.

[102] H. Hwang, S.-H. Kim, T.-H. Kim, J.-K. Park, and Y.-K. Cho. Paper on a disc: balancing the capillary-driven flow with a centrifugal force. *Lab on a Chip*, 11(20):3404–3406, 2011.

[103] N. Ichikawa, K. Hosokawa, and R. Maeda. Interface motion of capillary-driven flow in rectangular microchannel. *Journal of Colloid and Interface Science*, 280(1):155–164, 2004.

[104] B. Jacod, F. Pubilier, P. Cann, and A. Lubrecht. An analysis of track replenishment mechanisms in the starved regime. *Tribology Series*, 36:483–492, 1999.

[105] T.-J. Je, D.-S. Choi, E.-C. Jeon, E.-S. Park, and H.-J. Choi. Study on machining high aspect ratio micro channel structures using a shaping method by a diamond tool. In *Proceedings of the 27th American Society for Precision Engineering (ASPE) Annual Meeting*, volume 54, pages 433–436, 2012.

[106] L. Jiang and H.-L. Tsai. Prediction of crater shape in femtosecond laser ablation of dielectrics. *Journal of Physics D: Applied Physics*, 37(10):1492, 2004.

[107] V. Jokinen, M. Leinikka, and S. Franssila. Microstructured surfaces for directional wetting. *Advanced Materials*, 21(47):4835–4838, 2009.

[108] V. Jokinen, L. Sainiemi, and S. Franssila. Controlled lateral spreading and pinning of oil droplets based on topography and chemical patterning. *Langmuir*, 27(11):7314–7320, 2011.

[109] J. Judy, D. Maynes, and B. Webb. Characterization of frictional pressure drop for liquid flows through microchannels. *International Journal of Heat and Mass Transfer*, 45(17):3477–3489, 2002.

[110] S. G. Kandlikar, D. Schmitt, A. L. Carrano, and J. B. Taylor. Characterization of surface roughness effects on pressure drop in single-phase flow in minichannels. *Physics of Fluids*, 17(10):100606, 2005.

[111] K. Karki and S. Patankar. Calculation procedure for viscous incompressible flows in complex geometries. *Numerical Heat Transfer, Part A: Applications*, 14(3):295–307, 1988.

[112] G. Karniadakis, A. Beskok, and N. Aluru. *Microflows and Nanoflows: Fundamentals and Simulation*. Springer, 2005.

[113] D. S. Kim and T. H. Kwon. Modeling, analysis and design of centrifugal force-driven transient filling flow into a circular microchannel. *Microfluidics and Nanofluidics*, 2(2):125–140, 2006.

[114] K. S. Kim and M.-S. Chun. Pulsatile Poiseuille flows in microfluidic channels with back-and-forth mode. *Korea-Australia Rheology Journal*, 24(2):89–95, 2012.

[115] B. J. Kirby. *Micro- and nanoscale fluid mechanics: Transport in microfluidic devices*. Cambridge University Press, 2010.

[116] J. Klatte, D. Haake, M. Weislogel, and M. Dreyer. A fast numerical procedure for steady capillary flow in open channels. *Acta Mechanica*, 201(1):269–276, 2008.

[117] S. Klien, A. Diem, and C. Surberg. Untersuchungen zur temperaturabhängigen Schmierstoffmigration und Oberflächenspannungen verschiedener Werkstoffe. In *Tagungsband zum ÖTG-Symposium 2011*, pages 105–112, 2011.

[118] J. Klima, M. Hinterkausen, G. Dornhöfer, and W. Seemann. Oil transport in microchannels by means of capillary forces. In *Proceedings of the 42nd Leeds-Lyon Symposium on Tribology*, page 149, 2015.

[119] J. Koo and C. Kleinstreuer. Liquid flow in microchannels: experimental observations and computational analyses of microfluidics effects. *Journal of Micromechanics and Microengineering*, 13(5):568, 2003.

[120] K. Kotz, K. Noble, and G. Faris. Optical microfluidics. *Applied Physics Letters*, 85(13):2658–2660, 2004.

[121] H. Kusumaatmaja, C. Pooley, S. Girardo, D. Pisignano, and J. Yeomans. Capillary filling in patterned channels. *Physical Review E*, 77(6):067301, 2008.

[122] A. A. Kuznetsov and V. M. Martynov. Contact angle of lubricating oils at elevated temperatures. *Chemistry and Technology of Fuels and Oils*, 8(10):773–776, 1972.

[123] L. Landau and E. Lifshitz. *Fluid mechanics*, volume 6 of *Course of Theoretical Physics*. Elsevier, 2nd edition, 1987.

[124] D. W. Langbein. *Capillary surfaces: shape – stability – dynamics, in particular under weightlessness*. Springer, 2002.

[125] L. G. Leal. *Advanced transport phenomena: Fluid mechanics and convective transport processes*. Cambridge University Press, 2007.

[126] B.-K. Lee. Microinjection molding of plastic microfluidic chips including circular microchannels. *Polymer Engineering & Science*, 2013.

[127] H. J. Lee, D. J. Fermın, R. M. Corn, and H. H. Girault. Marangoni flow in micro-channels. *Electrochemistry Communications*, 1(5):190–193, 1999.

[128] R. Lenormand, C. Zarcone, and A. Sarr. Mechanisms of the displacement of one fluid by another in a network of capillary ducts. *Journal of Fluid Mechanics*, 135(34):337–353, 1983.

[129] W. W.-F. Leung and Y. Ren. Crossflow and mixing in obstructed and width-constricted rotating radial microchannel. *International Journal of Heat and Mass Transfer*, 64:457–467, 2013.

[130] W. W.-F. Leung and Y. Ren. Scale-up on mixing in rotating microchannel under subcritical and supercritical operating modes. *International Journal of Heat and Mass Transfer*, 77:157–172, 2014.

[131] H. Liang, D. Guo, L. Ma, and J. Luo. Experimental investigation of centrifugal effects on lubricant replenishment in the starved regime at high speeds. *Tribology Letters*, 59(1):1–9, 2015.

[132] M. Liu, J. Zhang, Y. Liu, W. Lau, and J. Yang. Modeling of flow burst, flow timing in Lab-on-a-CD systems and its application in digital chemical analysis. *Chemical Engineering & Technology*, 31(9):1328–1335, 2008.

[133] T. Lubrecht, D. Mazuyer, and P. Cann. Starved elastohydrodynamic lubrication theory: application to emulsions and greases. *Comptes Rendus de l'Académie des Sciences – Series IV*, 2(5):717–728, 2001.

[134] R. Lucas. Ueber das Zeitgesetz des kapillaren Aufstiegs von Flüssigkeiten. *Kolloid-Zeitschrift*, 23(1):15–22, 1918.

[135] V. Ludviksson and E. Lightfoot. The dynamics of thin liquid films in the presence of surface-tension gradients. *AIChE Journal*, 17(5):1166–1173, 1971.

[136] P. M. Lugt. *Grease lubrication in rolling bearings*. John Wiley & Sons, 2013.

[137] P. Man, C. Mastrangelo, M. Burns, and D. Burke. Microfabricated capillarity-driven stop valve and sample injector. In *Proceedings of the 11th Annual International Workshop on Micro Electro Mechanical Systems*, pages 45–50, 1998.

[138] T. Mang and W. Dresel. *Lubricants and lubrication*. John Wiley & Sons, 2007.

[139] J. Mann Jr, L. Romero, R. Rye, and F. Yost. Flow of simple liquids down narrow V grooves. *Physical Review E*, 52(4):3967, 1995.

[140] J. P. McHale and S. V. Garimella. Heat transfer in trapezoidal microchannels of various aspect ratios. *International Journal of Heat and Mass Transfer*, 53(1):365–375, 2010.

[141] C. D. Meinhart and H. Zhang. The flow structure inside a microfabricated inkjet printhead. *Journal of Microelectromechanical Systems*, 9(1):67–75, 2000.

[142] D. Mikhaylov. *Untersuchungen zur Schmierstoffversorgung tribologischer Kontakte mittels Oberflächengestaltung unter besonderer Berücksichtigung thermischer Einflüsse*. Master's thesis, RWTH Aachen, 2014.

[143] M. Mohammadpour, P. Johns-Rahnejat, H. Rahnejat, and R. Gohar. Boundary conditions for elastohydrodynamics of circular point contacts. *Tribology Letters*, 53(1):107–118, 2014.

[144] L. F. Moody. Friction factors for pipe flow. *Transactions of the ASME*, 66(8):671–684, 1944.

[145] A. Münch. The thickness of a Marangoni-driven thin liquid film emerging from a meniscus. *SIAM Journal on Applied Mathematics*, 62(6):2045–2063, 2002.

[146] R. Nilson, S. Tchikanda, S. Griffiths, and M. Martinez. Steady evaporating flow in rectangular microchannels. *International Journal of Heat and Mass Transfer*, 49(9):1603–1618, 2006.

[147] S. O'Brien and L. Schwartz. Theory and modeling of thin film flows. In A. T. Hubbard, editor, *Encyclopedia of Surface and Colloid Science*, volume 1, pages 5283–5297. CRC press, 2002.

[148] J. Ockendon. *Viscous flow*. Cambridge University Press, 1995.

[149] A. Oron, S. H. Davis, and S. G. Bankoff. Long-scale evolution of thin liquid films. *Reviews of Modern Physics*, 69(3):931 – 980, 1997.

[150] J. Ou and J. P. Rothstein. Direct velocity measurements of the flow past drag-reducing ultrahydrophobic surfaces. *Physics of Fluids*, 17(10):103606, 2005.

[151] F. F. Ouali, G. McHale, H. Javed, C. Trabi, N. J. Shirtcliffe, and M. I. Newton. Wetting considerations in capillary rise and imbibition in closed square tubes and open rectangular cross-section channels. *Microfluidics and Nanofluidics*, 15(3):309–326, 2013.

[152] K. Owen, B. VanDerElzen, R. Peterson, and K. Najafi. High aspect ratio deep silicon etching. In *Proceedings of the IEEE 25th International Conference on Micro Electro Mechanical Systems (MEMS)*, pages 251–254, 2012.

[153] L. Paouris, R. Rahmani, S. Theodossiades, H. Rahnejat, G. Hunt, and W. Barton. An analytical approach for prediction of elastohydrodynamic friction with inlet shear heating and starvation. *Tribology Letters*, 64(1):10, 2016.

[154] T. Patzek and D. Silin. Shape factor and hydraulic conductance in noncircular capillaries: I. One-phase creeping flow. *Journal of Colloid and Interface Science*, 236(2):295–304, 2001.

[155] G. Poll, X. M. Li, F. Guo, and N. Bader. Starved lubrication in rolling contact – a review. In *20th International Colloquium Tribology*. Technische Akademie Esslingen, 2016.

[156] A. Ponomarenko, D. Quéré, and C. Clanet. A universal law for capillary rise in corners. *Journal of Fluid Mechanics*, 666:146–154, 2011.

[157] V. Popov. *Contact mechanics and friction: Physical principles and applications*. Springer, 2nd edition, 2017.

[158] C. Powell, M. Savage, and P. Gaskell. Modelling the meniscus evacuation problem in direct gravure coating. *Chemical Engineering Research and Design*, 78(1):61–67, 2000.

[159] H. Princen. Capillary phenomena in assemblies of parallel cylinders: I. Capillary rise between two cylinders. *Journal of Colloid and Interface Science*, 30(1):69–75, 1969.

[160] S. Qian, D. Guo, S. Liu, and X. Lu. Experimental investigation of lubricant flow properties under micro oil supply condition. *Journal of Tribology*, 134(4):041501, 2012.

[161] M. Queralt-Martín, M. Pradas, R. Rodríguez-Trujillo, M. Arundell, E. C. Poiré, and A. Hernández-Machado. Pinning and avalanches in hydrophobic microchannels. *Physical Review Letters*, 106(19):194501, 2011.

[162] D. Quéré. Wetting and roughness. *Annual Review of Materials Research*, 38:71–99, 2008.

[163] T. Ransohoff and C. Radke. Laminar flow of a wetting liquid along the corners of a predominantly gas-occupied noncircular pore. *Journal of Colloid and Interface Science*, 121(2):392–401, 1988.

[164] A. Rawool, S. K. Mitra, and S. Kandlikar. Numerical simulation of flow through microchannels with designed roughness. *Microfluidics and Nanofluidics*, 2(3):215–221, 2005.

[165] Y. Ren and W. W.-F. Leung. Flow and mixing in rotating zigzag microchannel. *Chemical Engineering Journal*, 215:561–578, 2013.

[166] M. Reyssat, L. Courbin, E. Reyssat, and H. A. Stone. Imbibition in geometries with axial variations. *Journal of Fluid Mechanics*, 615:335–344, 2008.

[167] L. Romero and F. Yost. Flow in an open channel capillary. *Journal of Fluid Mechanics*, 322:109–129, 1996.

[168] P. Roy, N. Anand, and D. Banerjee. Numerical simulation of flow and heat transfer in radially rotating microchannels. *Microfluidics and Nanofluidics*, 15(3):397–413, 2013.

[169] P. Roy, N. Anand, and D. Banerjee. A review of flow and heat transfer in rotating microchannels. *Procedia Engineering*, 56:7–17, 2013.

[170] R. V. Roy, A. J. Roberts, and M. Simpson. A lubrication model of coating flows over a curved substrate in space. *Journal of Fluid Mechanics*, 454:235–261, 2002.

[171] R. Safavieh, A. Tamayol, and D. Juncker. Serpentine and leading-edge capillary pumps for microfluidic capillary systems. *Microfluidics and Nanofluidics*, 18(3):357–366, 2015.

[172] O. Saita. Evaluation of greases contributing to maintenance interval extension of shinkansen's traction motor. *NLGI Spokesman*, 73(5):38–45, 2009.

[173] T. S. Sammarco and M. A. Burns. Thermocapillary pumping of discrete drops in microfabricated analysis devices. *AIChE Journal*, 45(2):350–366, 1999.

[174] L. Schwartz and D. Weidner. Modeling of coating flows on curved surfaces. *Journal of Engineering Mathematics*, 29(1):91–103, 1995.

[175] T. Sochi. Newtonian flow in converging-diverging capillaries. *International Journal of Modeling, Simulation, and Scientific Computing*, 4(03):1350011, 2013.

[176] S. Solovyev. *Reibungs- und Temperaturberechnung an Festkörper- und Mischreibungskontakten.* PhD thesis, Otto-von-Guericke-Universität Magdeburg, 2006.

[177] J. Spurk and N. Aksel. *Fluid mechanics.* Springer, 2nd edition, 2008.

[178] T. M. Squires and S. R. Quake. Microfluidics: Fluid physics at the nanoliter scale. *Reviews of Modern Physics*, 77(3):977, 2005.

[179] M. E. Steinke and S. G. Kandlikar. Single-phase liquid friction factors in microchannels. *International Journal of Thermal Sciences*, 45(11):1073–1083, 2006.

[180] H. A. Stone, A. D. Stroock, and A. Ajdari. Engineering flows in small devices: Microfluidics toward a Lab-on-a-Chip. *Annual Review of Fluid Mechanics*, 36:381–411, 2004.

[181] A. D. Stroock, S. K. Dertinger, A. Ajdari, I. Mezić, H. A. Stone, and G. M. Whitesides. Chaotic mixer for microchannels. *Science*, 295(5555):647–651, 2002.

[182] J.-S. Suh, R. Greif, and C. P. Grigoropoulos. Friction in micro-channel flows of a liquid and vapor in trapezoidal and sinusoidal grooves. *International Journal of Heat and Mass Transfer*, 44(16):3103–3109, 2001.

[183] A. Z. Szeri. *Fluid film lubrication.* Cambridge University Press, 2nd edition, 2011.

[184] M. Taroni, C. Breward, P. Howell, and J. Oliver. Boundary conditions for free surface inlet and outlet problems. *Journal of Fluid Mechanics*, 708:100–110, 2012.

[185] M. Tatineni and X. Zhong. Numerical study of two-phase flows in microchannels using the level set method. In *Proceedings of the 42nd AAIA Aerospace Sciences Meeting and Exhibit*, page 929, 2004.

[186] J. B. Taylor, A. L. Carrano, and S. G. Kandlikar. Characterization of the effect of surface roughness and texture on fluid flow – past, present, and future. *International Journal of Thermal Sciences*, 45(10):962–968, 2006.

[187] S. Tchikanda, R. Nilson, and S. Griffiths. Modeling of pressure and shear-driven flows in open rectangular microchannels. *International Journal of Heat and Mass Transfer*, 47(3):527–538, 2004.

[188] G. F. Teletzke, H. T. Davis, and L. Scriven. How liquids spread on solids. *Chemical Engineering Communications*, 55(1-6):41–82, 1987.

[189] J.-t. Teng, C. Liu, C. Zhang, J.-C. Chu, J.-H. Cheng, M.-T. Lee, R. Greif, S. Jin, S. Huang, T. Dang, et al. Fluid dynamics in microchannels. In L. H. Juarez, editor, *Fluid Dynamics, Computational Modeling and Applications*, chapter 17, pages 403–436. InTech, 2012.

[190] S. K. Thomas, R. C. Lykins, and K. L. Yerkes. Fully developed laminar flow in trapezoidal grooves with shear stress at the liquid–vapor interface. *International Journal of Heat and Mass Transfer*, 44(18):3397–3412, 2001.

[191] N. Tipei. Boundary conditions of a viscous flow between surfaces with rolling and sliding motion. *Journal of Tribology*, 90(1):254–261, 1968.

[192] Y.-T. Tseng, F.-G. Tseng, Y.-F. Chen, and C.-C. Chieng. Fundamental studies on micro-droplet movement by Marangoni and capillary effects. *Sensors and Actuators A: Physical*, 114(2):292–301, 2004.

[193] K. Tsougeni, D. Papageorgiou, A. Tserepi, and E. Gogolides. "Smart" polymeric microfluidics fabricated by plasma processing: controlled wetting, capillary filling and hydrophobic valving. *Lab on a Chip*, 10(4):462–469, 2010.

[194] L. Ubbelohde and G. Höttner. *Zur Viskosimetrie*. S. Hirzel Verlag, 1965.

[195] E. van Emden, C. Venner, and G. Morales-Espejel. Aspects of flow and cavitation around an EHL contact. *Tribology International*, 95:435–448, 2016.

[196] M. van Zoelen. *Thin layer flow in rolling element bearings*. PhD thesis, University of Twente, 2009.

[197] M. van Zoelen, P. Lugt, and C. Venner. Free surface thin layer flow on bearing raceways. *Journal of Tribology*, 130(2):021802, 2008.

[198] S. J. Vanhook, M. F. Schatz, J. Swift, W. McCormick, and H. L. Swinney. Long-wavelength surface-tension-driven Bénard convection: experiment and theory. *Journal of Fluid Mechanics*, 345(1):45–78, 1997.

[199] C. E. Vest. Anti-creep films for oil lubricants. *NASA Materials Engineering Branch Technical Information Paper*, No. 068, 2002.

[200] D. von der Linde and K. Sokolowski-Tinten. The physical mechanisms of short-pulse laser ablation. *Applied Surface Science*, 154:1–10, 2000.

[201] R. Wahl. *Untersuchung des Einflusses von Mikrotexturierungen auf den einsinnigen, ölgeschmierten Gleitkontakt von Stahl-Saphir-Paarungen*. PhD thesis, Karlsruher Institut für Technologie, 2010.

[202] L. Wang and F. Liu. Forced convection in slightly curved microchannels. *International Journal of Heat and Mass Transfer*, 50(5):881–896, 2007.

[203] D. T. Wasan, A. D. Nikolov, and H. Brenner. Droplets speeding on surfaces. *Science*, 291(5504):605–606, 2001.

[204] E. W. Washburn. The dynamics of capillary flow. *Physical Review*, 17(3):273–283, 1921.

[205] L. D. Wedeven, D. Evans, and A. Cameron. Optical analysis of ball bearing starvation. *Journal of Tribology*, 93(3):349–361, 1971.

[206] S. J. Weinstein and K. J. Ruschak. Coating flows. *Annual Review of Fluid Mechanics*, 36:29–53, 2004.

[207] C. Weir and J. Hoffman. Compressibilities of long-chain normal hydrocarbons. *Journal of Research of the National Bureau of Standards*, 55(6):307–310, 1955.

[208] M. M. Weislogel. Compound capillary rise. *Journal of Fluid Mechanics*, 709:622–647, 2012.

[209] M. M. Weislogel, J. A. Baker, and R. M. Jenson. Quasi-steady capillarity-driven flows in slender containers with interior edges. *Journal of Fluid Mechanics*, 685:271–305, 2011.

[210] M. M. Weislogel, Y. Chen, and D. Bolleddula. A better nondimensionalization scheme for slender laminar flows: The Laplacian operator scaling method. *Physics of Fluids*, 20(9):093602, 2008.

[211] D. Xia, L. M. Johnson, and G. P. López. Anisotropic wetting surfaces with one-dimensional and directional structures: Fabrication approaches, wetting properties and potential applications. *Advanced Materials*, 24(10):1287–1302, 2012.

[212] L.-J. Yang, T.-J. Yao, and Y.-C. Tai. The marching velocity of the capillary meniscus in a microchannel. *Journal of Micromechanics and Microengineering*, 14(2):220, 2004.

[213] W.-B. Young. Simulation of the filling process in molding components with micro channels. *Microsystem Technologies*, 11(6):410–415, 2005.

[214] D. Yu, X. Hu, C. Guo, T. Wang, X. Xu, D. Tang, X. Nie, L. Hu, F. Gao, and T. Zhao. Investigation on meniscus shape and flow characteristics in open rectangular microgrooves heat sinks with micro-PIV. *Applied Thermal Engineering*, 61(2):716–727, 2013.

[215] S. Zahedi, K. Gustavsson, and G. Kreiss. A conservative level set method for contact line dynamics. *Journal of Computational Physics*, 228(17):6361–6375, 2009.

[216] S. Zehnle, F. Schwemmer, R. Bergmann, F. von Stetten, R. Zengerle, and N. Paust. Pneumatic siphon valving and switching in centrifugal microfluidics controlled by rotational frequency or rotational acceleration. *Microfluidics and Nanofluidics*, 19(6):1259–1269, 2015.

[217] L. Zhang. *Experimental approach to fluid flow in capillary surface structures in tribological context*. Master's thesis, Technische Universität Chemnitz, 2015.

[218] W. Zheng, L.-P. Wang, D. Or, V. Lazouskaya, and Y. Jin. Role of mixed boundaries on flow in open capillary channels with curved air–water interfaces. *Langmuir*, 28(35):12753–12761, 2012.

[219] X. Zheng and Z.-h. Silber-Li. Measurement of velocity profiles in a rectangular microchannel with aspect ratio $\alpha = 0.35$. *Experiments in Fluids*, 44(6):951–959, 2008.

[220] B. Zhmud, F. Tiberg, and K. Hallstensson. Dynamics of capillary rise. *Journal of Colloid and Interface Science*, 228(2):263–269, 2000.

[221] D. Zhu, T. Nanbu, N. Ren, Y. Yasuda, and Q. Wang. Model-based virtual surface texturing for concentrated conformal-contact lubrication. *Proceedings of the Institution of Mechanical Engineers, Part J: Journal of Engineering Tribology*, 224(8):685–696, 2010.

[222] M. Zimmermann, P. Hunziker, and E. Delamarche. Valves for autonomous capillary systems. *Microfluidics and Nanofluidics*, 5(3):395–402, 2008.

KARLSRUHER INSTITUT FÜR TECHNOLOGIE (KIT)
SCHRIFTENREIHE DES INSTITUTS FÜR TECHNISCHE MECHANIK (ITM)

ISSN 1614-3914

Band 1 **Marcus Simon**
Zur Stabilität dynamischer Systeme mit stochastischer Anregung. 2004
ISBN 3-937300-13-9

Band 2 **Clemens Reitze**
Closed Loop, Entwicklungsplattform für mechatronische Fahrdynamikregelsysteme. 2004
ISBN 3-937300-19-8

Band 3 **Martin Georg Cichon**
Zum Einfluß stochastischer Anregungen auf mechanische Systeme. 2006
ISBN 3-86644-003-0

Band 4 **Rainer Keppler**
Zur Modellierung und Simulation von Mehrkörpersystemen unter Berücksichtigung von Greifkontakt bei Robotern. 2007
ISBN 978-3-86644-092-0

Band 5 **Bernd Waltersberger**
Strukturdynamik mit ein- und zweiseitigen Bindungen aufgrund reibungsbehafteter Kontakte. 2007
ISBN 978-3-86644-153-8

Band 6 **Rüdiger Benz**
Fahrzeugsimulation zur Zuverlässigkeitsabsicherung von karosseriefesten Kfz-Komponenten. 2008
ISBN 978-3-86644-197-2

Band 7 **Pierre Barthels**
Zur Modellierung, dynamischen Simulation und Schwingungsunterdrückung bei nichtglatten, zeitvarianten Balkensystemen. 2008
ISBN 978-3-86644-217-7

Band 8 **Hartmut Hetzler**
Zur Stabilität von Systemen bewegter Kontinua mit Reibkontakten am Beispiel des Bremsenquietschens. 2008
ISBN 978-3-86644-229-0

Band 9 **Frank Dienerowitz**
Der Helixaktor – Zum Konzept eines vorverwundenen Biegeaktors. 2008
ISBN 978-3-86644-232-0

Band 10 **Christian Rudolf**
Piezoelektrische Self-sensing-Aktoren zur Korrektur statischer Verlagerungen. 2008
ISBN 978-3-86644-267-2

Band 11 **Günther Stelzner**
Zur Modellierung und Simulation biomechanischer Mehrkörpersysteme. 2009
ISBN 978-3-86644-340-2

Band 12 **Christian Wetzel**
Zur probabilistischen Betrachtung von Schienen- und Kraftfahrzeugsystemen unter zufälliger Windanregung. 2010
ISBN 978-3-86644-444-7

Band 13 **Wolfgang Stamm**
Modellierung und Simulation von Mehrkörpersystemen mit flächigen Reibkontakten. 2011
ISBN 978-3-86644-605-2

Band 14 **Felix Fritz**
Modellierung von Wälzlagern als generische Maschinenelemente einer Mehrkörpersimulation. 2011
ISBN 978-3-86644-667-0

Band 15 **Aydin Boyaci**
Zum Stabilitäts- und Bifurkationsverhalten hochtouriger Rotoren in Gleitlagern. 2012
ISBN 978-3-86644-780-6

Band 16 **Rugerri Toni Liong**
Application of the cohesive zone model to the analysis of rotors with a transverse crack. 2012
ISBN 978-3-86644-791-2

Band 17 **Ulrich Bittner**
Strukturakustische Optimierung von Axialkolbeneinheiten. Modellbildung, Validierung und Topologieoptimierung. 2013
ISBN 978-3-86644-938-1

Band 18 **Alexander Karmazin**
Time-efficient Simulation of Surface-excited Guided Lamb Wave Propagation in Composites. 2013
ISBN 978-3-86644-935-0

Band 19 **Heike Vogt**
Zum Einfluss von Fahrzeug- und Straßenparametern auf die Ausbildung von Straßenunebenheiten. 2013
ISBN 978-3-7315-0023-0

Band 20 **Laurent Ineichen**
Konzeptvergleich zur Bekämpfung der Torsionsschwingungen im Antriebsstrang eines Kraftfahrzeugs. 2013
ISBN 978-3-7315-0030-8

Band 21 **Sietze van Buuren**
Modeling and simulation of porous journal bearings in multibody systems. 2013
ISBN 978-3-7315-0084-1

Band 22 **Dominik Kern**
Neuartige Drehgelenke für reibungsarme Mechanismen. 2013
ISBN 978-3-7315-0103-9

Band 23 **Nicole Gaus**
Zur Ermittlung eines stochastischen Reibwerts und dessen Einfluss auf reibungserregte Schwingungen. 2013
ISBN 978-3-7315-0118-3

Band 24 **Fabian Bauer**
Optimierung der Energieeffizienz zweibeiniger Roboter durch elastische Kopplungen. 2014
ISBN 978-3-7315-0256-2

Band 25 **Benedikt Wiegert**
Nichtlineare Schwingungen von Systemen
mit elastohydrodynamischen Linienkontakten. 2015
ISBN 978-3-7315-0350-7

Band 26 **Arsenty Tikhomolov**
Analytische, numerische und messtechnische
Untersuchung der Dynamik von Fahrzeugkupplungen
am Beispiel des Trennproblems. 2015
ISBN 978-3-7315-0362-0

Band 27 **Daniel Maier**
On the Use of Model Order Reduction Techniques
for the Elastohydrodynamic Contact Problem. 2015
ISBN 978-3-7315-0369-9

Band 28 **Xiaoyu Zhang**
Crosswind stability of vehicles under nonstationary
wind excitation. 2015
ISBN 978-3-7315-0376-7

Band 29 **Jens Deppler**
Ein Beitrag zur viskoelastischen Modellierung nichtholonomer
Bindungsgleichungen. 2017
ISBN 978-3-7315-0548-8

Band 30 **Georg Jehle**
Zur Modellbildung und Simulation reibungserregter
Schwingungen in Pkw-Schaltgetrieben. 2017
ISBN 978-3-7315-0668-3

Band 31 **Joachim Klima**
Lubricant transport towards tribocontact in capillary
surface structures. 2018
ISBN 978-3-7315-0814-4

Die Bände sind unter www.ksp.kit.edu als PDF frei verfügbar oder als Druckausgabe zu bestellen.